Recent Praise for THURSDAY'S UNIVERSE

"Masterful...In spinning her fascinating tale, Bartusiak captures not only the celestial elegance of the universe, but also the intellectual elegance of those who ponder it. She writes with considerable wit, and an ear for the pithy quote and the telling detail."

Dennis Meredith
Air & Space

"Authoritative...lucid...an insider's perspective on the discipline.... Readers who puzzle over the latest news from out there but don't know quite where it all fits in will welcome *THURSDAY'S UNIVERSE*."

Laurence A. Marschall
The *New York Times*

"This is unquestionably the book to have if you want to catch up on what's been happening in astronomy the last few years. It is solid, entertaining, and informative. You'll like it."

Robert Burnham
Astronomy magazine

"Bartusiak demonstrates a competence and graceful style that make this one of the more enjoyable armchair astronomy chronicles."

Kirkus Reviews

"Marcia Bartusiak has set herself a formidable task: to encompass all major recent developments in astronomy and cosmology. She succeeds rather well."

Sky & Telescope

"Remarkable ability to clarify abstract technical developments with easy-to-understand analogies.... [Bartusiak] shows a taste and intuition for the field that is far beyond that of most other science writers."

David Schramm
Louis Block Professor
The University of Chicago

The *New York Times* cited THURSDAY'S UNIVERSE as one of the best science books of 1987. In addition, the Astronomical Society of the Pacific named it a 1987 Astronomy Book of the Year.

THURSDAY'S UNIVERSE

MARCIA BARTUSIAK

TEMPUS™

Tempus Books of Microsoft Press
A Division of Microsoft Corporation
16011 NE 36th Way, Box 97017, Redmond, Washington 98073-9717

Published by arrangement with Times Books, a division of Random House, Inc.

Library of Congress Cataloging in Publication Data
Bartusiak, Marcia, 1950-
Thursday's universe.
Reprint. Originally published: 1st ed. : New York : Times Books in association with Omni Books, c1986.
Bibliography: p.
Includes index.
1. Astronomy—Popular works. I. Title.
[QB44.2.B37 1988] 520 88-20028
ISBN 1-55615-153-5

Printed and bound in the United States of America.

1 2 3 4 5 6 7 8 9 HCHC 3 2 1 0 9 8

Distributed to the book trade in the United States by Harper & Row.

Distributed to the book trade in Canada by General Publishing Company, Ltd.

To my parents, who once explained to a curious seven-year-old why a pot appears to be suspended in the northern sky

CONTENTS

The Earth, poet Archibald MacLeish once wrote, is but "a small . . . planet . . . of a minor star . . . at the edge of an inconsiderable galaxy in the immeasurable distances of space." Within that vastness resides the beautiful Whirlpool galaxy and its small companion, located about 14 million light-years from the Milky Way. *Courtesy of National Optical Astronomy Observatories.*

INTRODUCTION

○ ○ ○

Monday's universe is fair of face,
Tuesday's universe is full of grace,
Wednesday's universe is full of woe,
Thursday's universe has far to go . . .

—Variation on "Monday's
child is fair of face . . ."

A hush fell over Palomar mountain at twilight. Yet the stillness was deceptive. As the sky darkened above the California summit, a three-man team of astronomers was readying the 200-inch Hale telescope, an impressive assembly of steel and glass, to peer out to the farthest reaches of the visible universe in search of primordial clouds of gas, the stuff out of which galaxies were forged.

Earthly clouds, a thin layer of high cirrus, delayed the observation for a few hours, giving Wallace Sargent, of the California Institute of Technology, the opportunity to stroll on the exterior catwalk that wraps around the middle of the twelve-story-high observatory dome. Later that evening, Sargent's attentions would be focused on computer monitors and photon-counting devices, the modern means of contemplating the universe. But, for a moment, he looked at the firmament as in days of old. "There's Cassiopeia," he said, pointing to a prominent W-shaped pattern of stars to the north. "I didn't know the constellations at all when I first started observing, because I was trained as a theoretical physicist. But now, I always take the time to look at the sky before work. I never seem to tire of its fascinations." It is an urge, the famous American astronomer Edwin Hubble once noted, that is older than history.

Of late, contemplating the heavens has become an even more humbling experience than ever. The view from Palomar and other observatories has taught us that this 8,000-mile-wide ball of rock upon which we live is a mere pebble, compared with the vastness of space-time and the multiplicity of its inhabitants. This perception, of course, was attained only after humanity progressed through several stages of inquiry and technological expertise. To early stargazers, the sky surrounding Earth was simply a fair-faced Monday's universe. They imagined they stood at the center of a cosmos with exiled princesses, fabled beasts, and mighty heroes looking down from above. In certain ancient societies, the celestial sphere was a vaulted roof, through which the fires of the gods flickered. Other cultures thought of each heavenly light as the soul of a departed one. From these early, creative attempts to understand the star-studded sky, a science was born.

By making surprisingly accurate measurements of such celestial phenomena as the lunar cycles, planetary motions, and eclipses, astronomers of antiquity became ever more sophisticated in their interpretations of the heavens. And each time that humankind's model of the cosmos was altered, concomitant tremors rumbled through other realms of science, as well as theology and politics. Copernicus, with his doctrine of an Earth revolving about the Sun, removed us from the center of the cosmos; Galileo, the first person to gaze at the heavens with a telescope, introduced us to a universe of unexpected change and intricacy; Newton, ushering in an age of reason, convinced us that motions everywhere, both in the heavens and on Earth, are described by the same set of physical laws. A Tuesday's universe, one which advances with clocklike precision, was upon us.

In 1825, the French mathematician and philosopher Auguste Comte stated that further progress in astronomy would be severely limited. "Men will never encompass in their conceptions the whole of the stars," he declared. The very distance of celestial bodies, Comte pointed out, prevented observers from speculating on a star's physical or chemical properties. Astronomers seemed destined to be no more than celestial librarians, spending their nights painstakingly cataloging the positions and motions of thousands of stars. But within four decades of Comte's pronouncement, the chemistry of the universe did become accessible; scientists realized that a star's spectrum, its composite colors or wavelengths of light, contained within it the story of that star's composition and temperature. In a way, each spectral line, resolved with an instrument called a spectroscope, carries a stellar confidence directly to Earth.

Even with this knowledge, astronomers were constrained for a while to play an elaborate game of blindman's bluff with the universe. With lenses or mirrors pointed heavenward, observers gathered the visible light rays emanating from the nighttime sky, unaware that the universe was revealing a vastly different picture of itself in other regions of the electromagnetic spectrum. The bits of red, green, and blue that entered optical telescopes and placed their marks on photographic plates suggested that the universe was a rather serene place. Hubble, a young legal-scholar-turned-astronomer, did pry open a cosmic can of worms in 1929 when he made the startling discovery that we live in an expanding universe, where billions of galaxies rush away from one another at tremendous speeds. But, all in all, the well-ordered stars and elegant spiraling galaxies still maintained a graceful and dignified composure.

"When I began working in astronomy, in 1933," Leo Goldberg of Harvard has stated, "astronomy was almost strictly an observational science with very little understanding of the physical meaning of observations. A few brave souls were just beginning to apply the new discoveries of atomic physics to astronomy, but their work was viewed with a certain amount of skepticism. Nuclear physics as a science hardly existed. . . . The universe was a quiet and peaceful place, slowly evolving and exhibiting little of the violent and explosive behavior that we now take for granted."

The seeds of change, though, had already been planted. In 1888, the German physicist Heinrich Hertz had generated 2-foot-long radio waves in his laboratory; seven years later, Wilhelm Roentgen discovered X rays. These and other findings proved that there were both longer waves of electromagnetic radiation (radio waves can stretch out for miles) and shorter waves (X rays measure less than a millionth of an inch across). It's as if visible light, waves that measure about a hundred-thousandth of an inch from peak to peak, are but one octave on a piano, with the entire electromagnetic spectrum encompassing many keyboards.

As soon as observers were able to effectively explore these other parts of the spectrum, beyond the narrow band visible to the human eye, a new and golden era for astronomy was at hand. A perpetual shower of infrared, ultraviolet, radio, and x-ray energy falls upon the Earth from space, and our celestial eyesight was extended dramatically in the second half of this century with the development of such disciplines as radio and infrared astronomy. The vista widened even more as x-ray satellites, gamma-ray telescopes, and ultraviolet space observatories rose far above Earth's obscuring blanket of air, which greatly absorbs these high-energy radiations. Meanwhile, increases in computer power have allowed theorists to simulate a variety of cosmic events—from supernova explosions to the creation of galaxies out of a smooth, primeval plasma—right in their laboratories. Astronomy, the seminal science, has been reincarnated in the process.

"The intervening years have seen astronomy expand like the universe itself," said Goldberg, "driven by a sequence of technological and scientific advances and by a tenfold growth in the number of astronomers." A national report on the state of astronomy and astrophysics in the 1980s stated that "in all of history, there have been only two periods in which our view of the universe has been revolutionized within a single human lifetime. The first occurred three and a half centuries ago at the time of Galileo; the second is now under way."

Thursday's Universe describes the new celestial landscape that has emerged during this second revolution and traces the historic roots from which the changes sprang. Each chapter deals with a separate facet of the developing frontier, from stellar formation, black holes, and the Milky Way, to galaxies, quasars, and the Big Bang. No branch of astronomy has remained unaffected by the latest upheaval; each has had to incorporate some astonishing new finding into its models, often discarding long-held theories in the process. Arrays of radio antennas, aimed at the most remote corners of the cosmos, have observed the centers of certain young galaxies somehow spewing the energy of a trillion suns out of a space no larger than our solar system. Closer in, within our own stellar neighborhood, x-ray telescopes detect ten-mile-wide balls of pure nuclear matter, the collapsed remnants of massive stars, spinning dozens of times a second. Meanwhile, in the dust- and gas-filled sectors of the Milky Way, a host of molecules emit distinctive radio songs, pointing astronomers to the birthplaces of new stars . . . and new solar systems.

By turning a page in this book, you expend far more energy than radio telescopes throughout the world collect daily from interstellar space. Yet astronomers can gather these and other radiations and weave from the murmurs celestial tales of epic proportions. In the heart of our galaxy, a whirlpool of agitated matter appears to be swirling around a tiny object, which apparently contains the mass of a few million suns —a supermassive black hole perhaps? And many galaxies, while carried outward with the universe's expansion, are vigorously interacting with their neighbors—swapping mass, colliding, and triggering stupendous bursts of star formation in one another. The beauty of Hubble's expanding universe remains, but now it is a splendor infused with titanic energies: a universe of wonder, as well as woe.

At the same time, in a new and fruitful partnership with particle physicists, cosmologists are pushing back the barriers that once obstructed our understanding of the universe's birth in a primordial explosion some 15 billion years ago. New insights into the ways in which the basic forces of nature are related to one another are allowing theorists to imagine what may have happened a mere trillionth of a trillionth of a trillionth of a second after the moment of genesis. Enigmas that perplexed cosmologists for decades are unraveling: Why the early universe was once so hot, how it became so big, and why it keeps expanding. The most daring theorists are even beginning to venture how there came to be *something* in the universe rather than *nothing,* the very crux of all cosmological concerns. It may be that the four dimensions of space and time in which we live ballooned from an infinitesimally small kernel comprised of *ten* dimensions. Such are the concerns of the new cosmologists.

Mysteries still abound. Astronomers have gradually come to suspect that each and every galaxy in the universe is but an inconsequential raft floating within a much vaster sea of dark, invisible matter. This unseen material emits not one glimmer of light, yet it could account for more than 90 percent of the universe's mass. It may be nothing more than a host of "Jupiters" and planetoids, too small to observe with earthbound telescopes. Or maybe, some surmise, this dark matter is composed of exotic elementary particles that have not yet been detected in particle accelerators.

The cosmos, once thought of as blandly homogeneous over very large scales, is also beginning to exhibit a wondrous texture. Galaxies appear to be arranged as if they sit on the surfaces of huge, nested bubbles, each bubble stretching up to several hundreds of millions of light-years* across. This frothiness could very well be a telltale relic from the dawn of creation. The bubbly structure will surely serve as an historical record, enabling theorists to better trace how galaxies, clusters of galaxies, and superclusters originated.

Astronomy, it should be stressed, is more than a collection of facts. It is also people, observers and theoreticians alike, intent on wresting the secrets of the heavens out of each photon of light and cosmic ray that happens to reach the Earth or a spaceborne detector. *Thursday's Universe* is also their story. Those who are quoted in the succeeding pages are obviously only a small sample of the legion of scientists devoted to the study of the universe. The women and men, whose voices fill each chapter, should be considered as representatives for a truly global community of researchers.

*A light-year, a convenient unit of measurement in astronomy, is the distance light, moving at 186,000 miles per second, travels in one year. One light-year equals nearly six trillion miles.

THURSDAY'S
UNIVERSE

Silhouetted against a background of glowing gas, the swirling edge of a vast dark cloud of interstellar matter has assumed the profile of a horse. It is a stellar nursery. Eventually the gas and dust in the Horsehead nebula will condense into stars of various sizes. *Courtesy of National Optical Astronomy Observatories.*

GASES TO GASES, DUST TO DUST

> We had the sky up there all speckled with stars and we used to lay on our backs and look at them, and discuss about whether they was made or only just happened. Jim he allowed they was made, but I allowed they happened; I judged it would have took too long to make so many. Jim said the moon could 'a' laid them; well, that looked kind of reasonable, so I didn't say nothing against it . . .
>
> —Mark Twain, *The Adventures of Huckleberry Finn*

Astronomers are solitary creatures. To catch them at work, you often have to navigate tortuously winding roadways up ragged, mile-high mountainsides. But for many years, at least one observatory on this planet defied that convention. This particular telescope held the curious distinction of being accessible via the New York City subway system. You could catch the Broadway local at Times Square, travel northward in its graffiti-filled cars up the west side of Manhattan, and get off at the Columbia University station. A short walk to the southeast corner of 120th Street and Broadway then brought you to one of the smallest radio-telescope observatories in the world.

Its tiny, white-painted dome, a mere 12 feet in width, resided fourteen stories above the streets of Manhattan atop Columbia's old, red-brick Pupin Physics Laboratory. The juxtaposition of the disparate architectures was quite engaging: A modern radio telescope aimed for the stars, surrounded by copper-lion gargoyles, green with age.

From October to May, as cold, dry air flowed down from Canada, the dome's doors slid open to reveal a dishlike antenna only 4 feet wide, not unlike the many TV satellite receivers scattered about the United States' countryside. During each winter observing season, a small army of professors and graduate students aimed the dish toward the sky, making sure there was enough of a tilt to miss the city's skyline and the ubiquitous water towers that dot the urban rooftops. Their round-the-clock scans traced a smooth arc from the Citicorp Building's famous triangular profile in the south to the George Washington Bridge across the Hudson in the north. It seemed an odd enterprise to undertake in the heart of the Big Apple, but "it's best to set up your astronomical operations in the 'center of the universe,' " the telescope's engineer and die-hard New Yorker Sam Palmer often remarked, tongue in cheek.

Patrick Thaddeus, the observatory's founding father, who was then with the Goddard Institute for Space Studies in New York City, chose an ambitious mission for the telescope: to map, not the stars themselves, but the regions *between* the stars, starting from the vicinity of our Sun —a middle-aged dwarf star in a remote suburb of the Milky Way— and working inward to the teeming galactic core, with the hope of disclosing a few secrets concerning stellar birth along the way.

It is a tricky enterprise, for a major fraction of the loose gaseous material along the galaxy's interstellar byways is barely possible to detect, even with sophisticated instruments. But certain kinds of atoms and molecules do make their presence known. After absorbing energy during their tumbles and collisions within the Milky Way, the molecules emit distinctive radio or infrared waves. Each type of molecule can be identified by a unique spectral signature, a special voice that rises above the dissonant chatter of the galaxy to reveal the temperature, composition, and structure of the interstellar medium. "It is a fingerprint of enormous specificity that allows us to know more about the distant stars and interstellar gas than what's a few miles beneath our feet," notes Thaddeus, one of the world's foremost "astrochemists." "It's obliging of nature to provide us with such good information."

• 5 •

GASES TO GASES, DUST TO DUST

The question that Thaddeus has been attempting to answer with his radio mapping, conducted since the 1970s, is the very same issue that Huck and Jim pondered on their raft trip down the Mississippi: How are stars born? Where are they being made amidst the speckled sky? It is certainly a fundamental question in astronomy, but a complete understanding of the mechanism eluded astronomers for a very long time. Of all the varied stages in a star's life, the first one—birth—has been perhaps the least understood. How does a tenuous clump of atoms and molecules, the specific target of the astrochemist, truly become dense enough to contract gravitationally into that glowing ball of gases that we call a star?

While the Columbia team searched for answers from the west side of Manhattan, similar radio surveys have gone on around the globe, from Chile, Japan, and Australia to the mountainous regions of Texas, California, Arizona, and Massachusetts. Moreover, during these attempts to unravel the riddle of star formation, with both ground-based instruments and space telescopes, many new features in the galactic landscape are disclosed. For example, most stars, astronomers are finding, click on their nuclear-powered lights within gigantic, nearly opaque clouds, which are strung out along our galaxy's spiral arms like hundreds of beads on a necklace. As late as 1970, the very existence of these immense objects was totally unknown. Smaller clouds of dust and gas, more noticeable because of their resplendent glow and previously viewed as the most promising sites for stellar birth, turn out to be merely the tips of these dark cosmic icebergs. Many of these discoveries were not sudden revelations, as was the highly publicized detection of the first neutron star but, rather, slow realizations that evolved as radio and infrared astronomers began to reveal the true makeup of the interstellar medium, where the story of a star's birth begins.

It is empty in the vast regions between the stars. So empty that interstellar travelers out for a stroll might scoop up only a few atoms of hydrogen and perhaps one microscopic grain of dust if they grabbed at the void with both hands. By comparison, a cup of air here on Earth contains more than a billion trillion molecules. This low yield of cosmic flotsam makes space a vacuum millions of times finer than the artificial "vacuums" arduously produced in terrestrial laboratories. Yet, in certain regions throughout our galaxy, this tenuous sea of dust and gas becomes more dense, collecting into wispy clouds that astronomers have dubbed *nebulae.* The thickest of these clouds appear as dark silhouettes when situated in front of a bright array of distant stars. Sir William Herschel, a German musician who emigrated to England in 1757 and later became the private astronomer to King George III, mistakenly called them *Löcher im Himmel,* or holes in the heavens, since they look like yawning gaps in space. But each thimbleful of space in those dim, black regions can contain hundreds of atoms of hydrogen, the simplest and most abundant material in the universe, as well as a smattering of helium, traces of other elements and compounds, and a rarefied sprinkling of microscopic dust grains.

Of course, to earthlings that is an extraordinary void; if we used hydrogen instead of oxygen for our respiratory functions, we'd need to inhale about half a million cubic miles of such a "cloud" to get one decent gulp of air. But by celestial standards, such a density could be considered as crowded as a beach on a sunny July weekend. When these clouds are heated by nearby stars, or simply reflect the stars' radiations, the twisting patches of gas appear to glow with an inner fire, creating astronomy's most photogenic subjects. But these clouds are not only beautiful; they play an important role in the life cycle of the heavens. Each nebula marks the location of a stellar nursery, where dust and gas condense over tens of millions of years to give birth to hot, new stars —some as small as a twelfth of the mass of our Sun, others up to sixty times more massive.

Just a hundred years ago, without an understanding of modern nuclear physics to guide them, astronomers could not be sure that stars were actively forming in the sky, though some observers did venture that the clumpy regions of gas and dust in our galaxy might be the sites of on-going star formation. Even Herschel, a self-taught astronomer and avid telescope builder, described one bright cloud as "an unformed firey mist, the chaotic material of future suns." But observing a star in the throes of birth has not been easy. Until recently, it was nearly impossible, for the shutters are tightly closed when optical astronomers aim their instruments toward a cosmic cradle. The dust and gas obscure our view. Even after a newly born star turns on the nuclear furnace deep within its core and begins to shine, it remains hidden for a while behind the thick veil of material from which it condensed.

• 7 •

GASES TO GASES, DUST TO DUST

Earlier in this century, astronomers despaired that they would ever be privy to the secrets of stellar birth. Cecilia Payne-Gaposchkin, who in 1925 was the first person to receive a Ph.D. for work done at the prestigious Harvard Observatory, remarked more than thirty years ago that "the man who presumes to talk of the evolution of the stars must needs be an optimist with a sense of humor." Around the same time, British astronomer Fred Hoyle wrote that "star-formation is a process that in its actual occurrence is irrevocably hidden from us because it occurs deep inside large clouds where the dust . . . prevents us from inspecting the process as it goes on." But Hoyle's use of the word "irrevocably" was premature, for new methods of observation being developed at the very moment Hoyle was writing that sentence would soon transform the study of stellar evolution. Astronomers were about to look at the universe with more penetrating "eyes": radio and infrared eyes that could directly scrutinize the galaxy's star-forming regions.

It was a bit of serendipity that first disclosed the tremendous potential in exploring the suprahuman regions of the electromagnetic spectrum. In the 1930s Karl Jansky, fresh out of college with a degree in physics and recently hired by Bell Telephone Laboratories, built a crude antenna amid central New Jersey's potato fields to investigate strange hissing noises that were disrupting transatlantic radio-telephone communications. After a year of detective work with his spindly network of brass pipes hung over a wooden frame that rolled around on Model-T Ford wheels, Jansky established that the source of the 20-megahertz signal (a frequency between the United States' AM and FM bands) was neither in the Earth's atmosphere nor in the solar system, but came instead from the center of our home galaxy, the Milky Way. What Jansky affectionately called his "star noise" hinted at processes going on in the galactic nucleus that were not revealed by visible light rays emanating from that region. This cosmic static was composed of electromagnetic waves millions of times longer than the waves our eyes perceive as the color red.

Jansky didn't know it at the time, but some of the signals he picked up were being emitted by violent streams of charged particles spiraling about in the magnetic fields of space. Just as an oscillating electric current within a broadcast antenna releases waves of radio energy into the air, these energetic particles also broadcast radio waves, and Jansky, who did not start out to make an astronomical observation, was the first to detect them. Astronomy's reliance on the optical window was finally broken. We were no longer limited to peering into space; we could eavesdrop as well.

After World War II, spurred by the military development of radar technology, the infant field of radio astronomy burgeoned. Antennas sprouted around the globe, particularly in Great Britain, Australia, and the Netherlands. Many of the war's radar experts, eager to decipher the universe's varied radio cries, quickly became involved in the new effort. The technique was particularly useful to those hoping to catch a star-in-the-making. Unlike visible light, radio waves generated within a dense clump of gas and dust can freely pass through the debris without being absorbed, in the manner of radar passing through a fog. Such a propitious capability offered astronomers their first passkey to those seemingly impenetrable interstellar clouds.

Even before the war ended, a young Dutch student named Hendrik van de Hulst predicted that the single atoms of hydrogen floating in space (each atom consisting of one electron orbiting around a single proton) would be emitting a specific type of star noise, an unvarying radio "song." This would happen, he reasoned from atomic theory, when the lone spinning electron in a hydrogen atom occasionally flipped over to a lower energy state, generating an electromagnetic wave 21 centimeters (~8 inches) in length. Any one hydrogen atom out in space performs this flip on average every 11 million years or so, but van de Hulst pointed out that the enormous number of hydrogen atoms in our galaxy should collectively produce a continuous, detectable hum. And by 1951, when equipment at last became available to detect the weak signal, hydrogen's monotonic voice was recorded.

Observing 21-centimeter emissions across the sky has proved invaluable in mapping our galaxy's spiral arms, where atomic hydrogen tends to be abundant. And since isolated hydrogen atoms were also assumed to form the bulk of the denser interstellar clouds, astronomers were sure the atoms' radiations would provide them entry. But, to their vexation, the signal did not become stronger when they aimed their radio antennas toward the darkest, dustiest clumps. Indeed, sometimes it was weaker.

It turns out that the gas and dust were becoming sufficiently dense and the temperature low enough (around −440 degrees Fahrenheit, nearly 400 degrees colder than a typical day in Antarctica) for two individual hydrogen atoms to join up and form a molecule of hydrogen (H_2), which unfortunately does not emit a radio signal. So with no radio waves being sent out by H_2, radio astronomers were as blind as their optical counterparts when it came to examining the inner sanctum of a dark galactic cloud. It looked as if the breeding ground for stars would remain as impenetrable as ever.

But starting in the 1960s, astronomers began to detect the radio emissions from a host of other compounds that formed side by side with the molecular hydrogen. This was one of the more startling discoveries in recent astronomical history. Half a century ago, optical astronomers had detected a few molecules in space, such as the free chemical radicals carbon-hydrogen (abbreviated CH in chemical notation) and carbon-nitrogen (CN). The minute traces of CH and CN were measured when wisps of gas containing the molecules happened to be caught in front of a bright star, allowing astronomers to see the molecules absorb the starlight at certain wavelengths. Yet astronomers were firmly convinced that more complicated molecules could not survive the rigors of interstellar space; the molecules would be broken apart by energetic ultraviolet or cosmic rays, it was thought, even as they formed. In retrospect, that reasoning was quite faulty. In regions where densities of gas and dust are quite high, stray radiations that could tear complex molecules apart are hindered.

There are several ways a molecule in an opaque cloud can absorb and re-emit a bit of radiation (letting an astronomer know where it is), but one process dominates. Molecules often get kicked up to a higher spin, after randomly colliding with the many molecules of hydrogen in the cloud. Most of the excited molecules then send out their unique radio signals whenever they drop from the high rate of rotation to a lower, less energetic state.

By carefully monitoring these signals, radio astronomers have been able to detect more than sixty chemicals in space. The ever-growing list includes ammonia (NH_3), the welding fuel acetylene (C_2H_2), the preservative formaldehyde (H_2CO), and a bevy of exotic molecules too volatile to exist on Earth. In turbulent regions near newborn stars some molecules, such as water (H_2O) and hydroxyl (OH), even coalesce to form gigantic natural masers, the microwave equivalent of lasers. The molecules in these masers dump their energy simultaneously, to produce an intense beam of single-frequency radio energy, all the waves precisely in phase and reinforcing each other.*

The large majority of the molecules are combinations of hydrogen, carbon, nitrogen, and/or oxygen. The hydrogen was forged during the Big Bang, that conflagration that gave birth to our universe some fifteen eons (15 billion years) ago. But the weightier elements in these molecules were brewed deep inside stars and spewed into space, either gradually through the action of a stellar wind blowing off a star or catastrophically in a supernova explosion. The calcium in our bones, the iron in our blood, in fact each atom heavier than hydrogen in our bodies, can trace its origins to the heart of a star. One wonders if Darwin himself would have nodded in sage agreement upon hearing that stars are our ultimate ancestors.

Pure hydrogen, however, is still the overwhelming cosmic ingredient in an interstellar cloud. Only one ammonia molecule forms, for example, for every 30 million molecules of hydrogen. But taken over the vastness of space, even such trace amounts add up. The discoverers of celestial ethyl alcohol once noted with amusement that the galactic center contained enough alcohol to fill up more than a billion billion billion bottles of whiskey. Of course, you'd have to distill a volume as big as the planet Jupiter to mix a decent drink.

All this tells us that the Milky Way is nature's largest chemical laboratory. And what has surprised many is the propensity of galactic clouds to form organic (that is, carbon-based) compounds instead of exotic inorganic structures composed of silicon or metals. As Thaddeus, who has helped discover more than a quarter of the molecules known to exist in space, likes to put it, "Organic compounds are not just the scum on a planet's surface." A few astrochemists have even speculated that comets and carbon-rich meteorites (remnants of the solar system's birth from a whirling pool of gas) may have helped seed our planet with the simplest carbon compounds, the basic materials for such building blocks of life as amino acids and proteins.

The field of astrochemistry truly blossomed in 1970 when Robert

*In 1964, physicist Charles Townes was awarded the Nobel Prize for constructing the first maser (*m*icrowave *a*mplification by *s*timulated *e*mission of *r*adiation) in a Columbia University laboratory. The Nobel committee didn't realize that nature had preceded Townes by billions of years.

Wilson, Keith Jefferts, and Arno Penzias of Bell Telephone Laboratories attached a newly developed receiver to the National Radio Astronomy Observatory's radio telescope atop Kitt Peak in southern Arizona and detected the 2.6-millimeter (1/10 of an inch) electromagnetic waves being emitted by carbon monoxide (one atom of carbon joined to one atom of oxygen). It's a poisonous car emission here on Earth, but an invaluable guide to astronomers when found in the immense vacuum between the stars. Thaddeus likens it to a biological stain. "You can't see a nucleic acid or protein within a cell, so you have to use a drop of dye to bring out the structure. Well, in the densest star-forming regions, we're caught in a similar situation. We can't see the dominant molecule—molecular hydrogen—either." Traces of carbon monoxide do, however, mix in with the hydrogen, and their strong radio emissions act as a road map to the hidden topography of interstellar space. With the discovery of interstellar CO, radio astronomers obtained their "Holy Grail": Its stability and relative abundance made it an effective tracer of the invisible clouds of molecular hydrogen that were certain to be cooking up stars within their depths. By tracking carbon monoxide emissions, astronomers were able to start determining the density, temperature, and motions of these "molecular" clouds, as they came to be called. "How far do they extend, and what can they tell us about star formation?" observers began asking.

Early on, radio astronomers Philip Solomon and Nicholas Scoville concentrated their efforts on Jansky's old haunt, the center of the Milky Way in the direction of the constellation Sagittarius, and mapped clouds of molecules that contained the mass of more than a million suns along a stretch of space 800 to 900 light-years wide; more kinds of interstellar molecules have been detected within the depths of these clouds than in any other region of our galaxy. Later, with more extensive surveys, they saw that a considerable fraction of the gas in the galaxy's center is locked up in molecular clouds.

Meanwhile, Thaddeus and two colleagues, Kenneth Tucker and Marc Kutner, used a gold-plated 16-foot-wide radio dish at the McDonald Observatory in the hills of western Texas to scan a region in the Orion constellation, the hourglass figure which dominates the winter sky in the northern hemisphere. Orion was a logical choice. This sector abounds with bright gaseous clouds, and astronomers at the time believed that such nebulae marked the cores of star-forming regions. Their starting point was a patch of dust in Orion's belt that is known as the Horsehead nebula, because of its famous equine profile. "We worked outward from there, gradually aiming the telescope farther and farther over," says Thaddeus. "We 'felt the elephant,' so to speak, and ended up getting a feel for the scale of the beast."

The beast was sizable. After fifty hours of observation, sampling some 150 positions, they mapped a cloud 100 light-years across that contained enough matter to form 100,000 suns. But they later learned that this giant cloud was merely a branch of a much larger complex that stretches the entire length of Orion's torso. "It would have taken us years to cover the full extent with the McDonald telescope," notes Thaddeus. "It was like trying to paint a barn with a quarter-inch brush," or trying to photograph a panorama of the Grand Canyon with a telescopic lens: possible but not practical. This led to Thaddeus's decision to build a "larger" brush: his paradoxically tiny Manhattan-based radio telescope, which became an invaluable tool in figuring out some of the mysteries of star formation. The diminutive antenna began perusing the sky above Cambridge, Massachusetts, in 1986 when Thaddeus moved his operations to the Harvard-Smithsonian Center for Astrophysics, interestingly enough just a few blocks north of the Harvard stop on Boston's subway system.

As with any telescope, radio or optical, the smaller the aperture, the wider the view. Therefore, by going from a 16-foot-wide radio dish to one only 4 feet across, Thaddeus was acquiring a sort of wide-angle lens that could examine a larger swath of sky with each observation. Thaddeus's antenna is, in many ways, like a radio stuck on one station. With a specially designed receiver that's cooled to a chilly −456° Fahrenheit (just four degrees away from absolute zero, that unattainable temperature where, theoretically, all molecular motion would cease), the telescope is tuned to the piercing 115-gigahertz (115 billion cycles per second) cry of celestial carbon monoxide, a signal that travels easily through our atmosphere without getting absorbed. That's a frequency appreciably higher than the waves in a microwave oven or the many commercial radio and television signals crowding city airwaves. "So, to us, the New York environment was as quiet as it was when Henry Hudson came up the river," says Thaddeus.

Throughout the 1970s and early 1980s, surveys conducted by Thaddeus and others confirmed that a sizable portion of the interstellar gas* —perhaps more than one-half—is confined within some kind of cold molecular cloud. Like a forest with its varied grasses and trees, the clouds come in a range of sizes. The smallest contain less than 100 solar masses and are the most numerous. These are Herschel's *Löcher im Himmel.* However, most of the gas is locked up in thousands of giant molecular clouds that can stretch as much as 300 light-years across (about 300,000 of our solar systems set end to end) and contain the mass of hundreds of thousands, even millions of suns. This makes them the most massive objects in the Milky Way, and hence the most prolific stellar delivery rooms, yet two decades ago no one even imagined that they existed. Some astronomers suspect that the larger clouds congregate along our galaxy's spiral arms like pearls on a string.

Were the solar system to pass through one of these giant clouds, the journey would be dismal indeed. Astronomers Scoville and Judith Young once painted a bleak portrait of the passage: "The absorption of light would be so great that all but the few nearest stars would disappear from view. . . . More than two million years would elapse before the earth emerged from the gloom. Given the abundance of the clouds in our galaxy, such an event must come roughly once every billion years. . . . If human beings had evolved during such an episode, both their vision of the universe and their philosophical outlook would have been fundamentally different."

Fortunately for astronomy and the many poets who have written odes to the star-filled heavens, our Sun is presently caught in the center of a pristine bubble of interstellar space that may have been cleared out by a series of supernova explosions that detonated about 60 million years ago. The Great Rift of Cygnus, the dark lanes of gas and dust 400 light-years distant that on a clear summer night can be seen to split the Milky Way in half, may well be some fragments blown off the molecular cloud that had earlier given birth to the exploding stars.

While giant molecular clouds were being mapped and cataloged, one question remained unanswered: How were stars being ignited within

*It is estimated that interstellar gases comprise 5 to 10 percent of the mass of our galaxy; microscopic dust grains composed of silicon, graphite, and/or various ices make up another 1 percent. Most of the mass is tied up in stars.

their depths? Although there were plenty of tentative conjectures, the overall picture was unfocused until the mid-1970s, when a graduate student at Harvard University named Charles Lada stumbled upon a compelling piece of evidence while searching for clouds of carbon monoxide near regions of bright, highly energized hydrogen.

One of the objects he examined closely was the Omega nebula, located in a particularly rich sector of the sky. The nebula's brilliant glow is caused by the hottest and most massive stars embedded in the cloud. These bright-blue O and B stars* emit copious amounts of ultraviolet radiation that ionizes much of the gas (that is, strips electrons off the atoms). When the electrons and ions recombine, excess energy is radiated outward, making the nebula visible. There is red fluorescence from hydrogen, sulfur, and nitrogen, as well as a delicate greenish tint from the oxygen atoms scattered about the gaseous filaments. Nebulae are no less than cosmic "neon signs."

When Lada aimed a radio telescope toward Omega, he expected to find the carbon monoxide molecules neatly dispersed around the entire nebula like some kind of chocolate icing. He logically assumed, as many did, that a nebula formed within the deepest recesses of a dark molecular cloud. But Lada didn't see that at all. Instead, the CO molecules were concentrated in a long dark swath that swept off to one side of the nebula. This clinched the idea, already suspected by some, that every bright nebula is merely a blister perched on the side of a molecular cloud, with the nebula's hot, young progeny eating away at the cloud's surface like some festering wound.

Yet, there was something more: There was a definitive sequence in the apparent evolution of stars in this region. First, off to one side of the bright Omega cloud, Lada noticed an older cluster of O and B stars, whose dusty blanket had long blown away; all that remained were the sharp pinpoints of light. Next in line was the nebula itself, which houses a younger set of stars that cannot yet be seen with optical telescopes but readily appear as "hotspots" with sensitive infrared instruments. Omega's thick curtain of dust and gas is absorbing the stars' visible light

*Stars are classified according to their spectral type and temperature. When arranged in order of decreasing temperature, the classes are O B A F G K M R N S. Astronomy students remember the sequence with the (somewhat unliberated) refrain, "Oh, Be A Fine Girl, Kiss Me. Right Now, Smack." Our yellow Sun is a G Star.

and re-emitting it at the longer infrared wavelength. Finally beyond this, there was the cold, seemingly starless mass of the molecular cloud itself. The arrow of time, Lada could see, was pointed toward the giant molecular cloud: The tremendous energies released by one generation of stars were apparently kindling a wave of star formation that surged down the length of the cloud like the successive bursts of a string of firecrackers.

The idea that one generation of stars can spark the birth of the next was not a new proposal; similar ideas had been bandied about for years. In 1964, for instance, the Dutch astronomer Adriaan Blaauw remarked that groups of O and B stars often line up in a certain way: the old, loose clusters at one end and the youngest and most compact groups at the other. Was this a family portrait, he asked? The German-American astronomer Walter Baade admitted that he got the distinct impression from his observations that star formation was spreading "in some way like a disease." Lada's molecular mapping provided an intriguing diagnosis.

This idea was better developed once astrophysicist Bruce Elmegreen arrived at Harvard in the fall of 1975 as a junior fellow. After looking at Lada's picture of the Omega neighborhood, Elmegreen immediately sensed the beauty of the scenario and worked out the theoretical details. "The intense radiations produced by the first generation of O and B stars are heating up the surface of the molecular cloud to temperatures around ten thousand degrees, which drives a shock wave into its depths," explains Elmegreen, who went on to IBM's Thomas J. Watson Research Center in New York. "This shock scoops up the gas like a snowplow, creating dense clumps that soon contract and ignite into the next generation of stars."

Eventually, these infant stars will blow away their placental covers and emerge at the surface of the cloud, only to continue the process as their own powerful energies dig ever deeper into the molecular cloud. One complete cycle can take fewer than several million years. Generation after generation can be spawned in this way, until the cloud is ultimately disrupted under the constant battering. Only a small percentage of the cloud's mass is actually converted into stars; the rest returns to interstellar space awaiting a wave of compression to forge yet another dense cloud.

The Elmegreen-Lada model has provided a useful framework for understanding the broad outline of star formation within the Milky Way and other galaxies, but astronomers are agreed that other mechanisms must also play a role in sparking the birth of a star. "This sequential process, of course, doesn't work everywhere," points out Lada, who is now with the University of Arizona. "If you have a chain reaction, you have to light the chain somehow. It's the old chicken-or-the-egg problem." A molecular cloud, Lada and others are learning, is a cauldron seething with varied shocks and pressures—each a potential trigger for star creation. When clouds themselves collide, for example, the compression of gas at the point of impact might lead to the production of protostars.

The more massive the star, the shorter is its life span. The rare and powerful O stars, stars 100,000 times as bright as our Sun and dozens of times as big, are cosmic spendthrifts that burn up all their fuel within three million years, a mere blink on the astronomical time scale (our own Sun is expected to shine 3,000 times longer). Therefore, some O stars can be expected to explode as glorious supernovae while still near the cloud that gave them birth. The jolt is sure to send out a shock wave powerful enough to compress the interstellar medium and form new stars. As evidence, astronomers William Herbst and George Assousa have detected swarms of young stars at the edges of the shells of gas swept out by ancient supernovae. It is a singularly satisfying cycle: new stars arising out of the death throes of a previous generation.

Stars, at least the most massive, do not emerge meekly into this universe, but rather gestate within an environment filled with turbulence. Once a star ignites, it fuels this cosmic hurricane with its own stellar winds, which can reach speeds of hundreds of miles per second. This has been one of the more surprising findings to come out of modern astronomy's snooping into the inner workings of a molecular cloud: Large young stars shed tremendous amounts of matter right after their birth. While the Sun emits 10^{-14} (one hundredth of a trillionth, or 0.00000000000001) of its mass each year as a solar wind, nascent blue-white giants can eject annually as much as a thousandth of a solar mass, at least for a while. That's about two trillion trillion tons of material (the mass of 300 planet Earths) per year. Theorists are hard pressed to understand why.

Much of the proof for this violent chapter in the evolution of a massive star comes from the Orion nebula. This hotbed of stellar activity is not the largest star-forming region in our galaxy, but at a distance of 1,600 light-years it is one of the closest, practically next door for astronomers.

This "Rosetta stone" on stellar birth is easy to locate in the nighttime sky. It's situated in the middle of the dagger that juts southward just below Orion's three-starred belt. This jewel in Orion's sword gets its glitter from the four young O and B stars blazing within its center. Together, they are called the Trapezium cluster and at one million years of age are the newest stars to cast off the dusty cloak that once hid them. The real wealth of information to be found in Orion, however, lies in the cold, dark molecular cloud located directly behind the nebula. While the Trapezium stars enter their adulthood on the surface of the cloud, a new generation of stars is already emerging just beneath. Carbon monoxide surveys readily point to the cache: CO emission from this region peaks sharply over an area a few tenths of a light-year across. Here, densities begin to exceed 100,000 molecules per cubic centimeter —conditions ripe for gravity to start squeezing the gas into stars. Though radio images of this area are fuzzy, higher-resolution infrared observations conveniently lift the veil.

For many years, the infrared section of the electromagnetic spectrum seemed destined to remain astronomy's forgotten child. The problem was one of detection. The human eye and photographic plates have served well in capturing visible light waves; antennas and receivers have taken care of viewing the radio wavelengths, at the other end of the spectrum. But infrared, or heat, radiation—that middle range of wavelengths extending from one millimeter (0.04 inch) down to a thousandth of a millimeter (0.00004 inch)—could not be gathered adequately by either technique.

Inventor Thomas Edison did devise an infrared detector in the 1870s. He observed a solar eclipse with it from a Wyoming henhouse and talked of mapping a hidden universe. This tantalizing window was never effectively opened, though, until the early 1960s when University of Arizona physicist Frank Low, then with Texas Instruments, built an infrared detector sensitive enough for useful astronomical work. Using liquid helium, he cooled the instrument to near absolute zero, making it responsive to the faint emanations of cosmic heat that fell on it at the focus of a reflecting telescope. Without such cooling, the detector would get swamped by the heat of its surroundings and thus "see" nothing.

Astronomical infrared detectors have now been installed on mountaintops, lofted in balloons, and flown on jets, mainly to get above atmospheric water vapor, which greatly absorbs infrared radiation. An extinct volcano named Mauna Kea on Hawaii has become infrared astronomy's mecca. Its summit, a stark, lava-strewn landscape reminiscent of the Viking photographs of Mars, reaches nearly three miles above sea level. This enables the special infrared observatories built on Mauna Kea's ancient cinder cones to stand above half the earth's atmosphere and ninety percent of its water vapor. What the Mauna Kea telescopes are seeing directly behind the Orion nebula is an assembly of infrared sources that collectively emit the energy of 100,000 of our Suns.

Some of the hotspots may be just clumps of gas shining by reflected light, but at least two of the sources, IRc2 and what's called the BN object,* seem to be gaseous orbs that have at last halted their gravitational contraction and ignited their thermonuclear engines. Hydrogen now fuses into helium deep within their hearts, releasing appreciable amounts of energy in the process. This occurs when the temperature at a star's center reaches over 10 million degrees. The ignition may even have occurred relatively recently, perhaps while Julius Caesar reigned over his Empire some 2,000 years ago.

As if by design, the universe even provides some trumpets to announce the birth: the cosmic masers. These pure beacons of microwave energy appear to be directly linked to an O or B star's debut. The process is initiated when the surface of the blue-white giant exceeds a temperature of 25,000 degrees Fahrenheit and begins to send out a torrent of atomic particles and ultraviolet photons. This intense stellar wind continually batters the placental cloud that still surrounds the infant star, pushing the gas farther and farther back. It literally blows a bubble in space.

Clumps of molecules in this expanding shell of gas, each clump as big as the Earth's orbit around the sun, absorb the energy from this onslaught and reradiate it as powerful microwave beams. Puzzled by the

*The first hotspot detected in Orion in 1965, named after its discoverers Eric Becklin and Gerry Neugebauer. IRc2 stands for infrared compact source number 2.

immense strength and pure frequency of these beams, the astronomers who first detected the signals once dubbed the then-unknown source "mysterium." By a process not yet completely understood, the molecules are kicked up to a single energy level and remain there until a chance event—the intrusion of a stray photon or the sudden decay of one of the pumped-up molecules—triggers the microwave avalanche. In this way the molecules dump their energy simultaneously, the hallmark of a maser signal. This cycle of absorption and emission is repeated continually. Microwave beacons have been found operating at frequencies belonging to water vapor, hydroxyl, methanol, and silicon monoxide.

Water masers are especially impressive. At any one time, dozens of them can surround a newly born star. Each individual water maser has a lifetime of a few years, but when one dies off another usually emerges to take its place. James Moran, a radio astronomer with the Harvard-Smithsonian Center for Astrophysics in Cambridge, Massachusetts, who has been studying masers for almost two decades, compares them to "fireflies that flicker on and off." This flickering may go on for 100,000 years, ending when the star's energetic photons finally rip through the placental cloud, ionizing everything in their path. The energies involved are not puny. There's one water maser in Orion that's the equivalent of a hundred-trillion-trillion-watt radio station! "The joke around here is that you hardly need an antenna at all to hear it, just a diode and an earplug," says Moran.

The maser group based at the Center for Astrophysics has been tracking the motions of water masers with an intercontinental array of radio telescopes that stretches from California to West Germany. By simultaneously collecting the masers' signals at antennas spaced over several continents, they simulate a single radio telescope as big as the Earth. This enables them to achieve resolutions a thousand times better than ground-based optical telescopes. With such sharp eyesight, Moran and his colleagues Mark Reid and Matthew Schneps have seen compact cloudlets of masing gas bulldozed away from newborn stars, such as IRc2 in Orion, at many miles per second. It was an early hint to astronomers that huge amounts of material flow out of young, massive stars, with the masers riding outward on the shock wave like cosmic surfers. The more massive the newborn star, the more pronounced the weight loss.

The spectacular fireworks associated with the appearance of massive stars have often overshadowed a less dramatic birthing process. IRc2 is not a typical character in the saga of star formation. Most of the stars inhabiting the Milky Way are actually mundane dwarfs, like our own Sun. How did they emerge, to use the words of Herschel, out of the "unformed firey mist?"

"This is a much more difficult process to observe, because low-mass stars are relatively faint. You can see them only in the very nearest regions," points out radio and infrared astronomer Philip Myers of the Harvard-Smithsonian Center for Astrophysics. Myers is one of a growing number of astronomers who are making low-mass star formation their specialty. They stay away from the hubbub of the giant Orion complex, preferring to study the small dark clouds located within a few dozen light-years of Earth. One of these is the Taurus cloud, a seemingly dormant stream of dust and gas that cuts across the Taurus constellation.

At times, ammonia acts as Myers's star-finder. In space, ammonia is a trace molecule that collects in only the densest interstellar regions. Using radio telescopes in Massachusetts and West Virginia to map the celestial microwave emissions of the noxious chemical, Myers discovered that the Taurus cloud was peppered with spots of ammonia, each about a quarter of a light-year across. Intriguingly, the total mass contained in each clump was enough to construct a star similar to our Sun. "And we noticed a surprising thing about these clumps of ammonia," says Myers. "In contrast to the turbulence found in giant molecular clouds, these dense cores are little oases of calm. They're dark, cool, and quiet. At the time, it made me wonder whether we were on the wrong track in looking for star-forming regions because I had the distinct impression that any region that forms a star would be fairly violent." Instead, the low-mass stars seemed to be condensing relatively quietly, like a gentle mist. A space telescope named IRAS would go Myers one better.

In January 1983, the Infrared Astronomical Satellite, a joint project of the Netherlands, Great Britain, and the United States, was launched from Vandenberg Air Force Base in California into a 560-mile-high orbit far above the Earth's disruptive lights and atmospheric moisture. Its mission ended ten months later after the satellite exhausted its vital supply of liquid helium, but not before the superchilled telescope revealed a universe vastly different from the one portrayed in optical photographs.

IRAS's primary mission was to survey the entire celestial sphere and catalog the heat radiation given off by the universe's myriad inhabitants; its detectors were so sensitive that they could have spotted the heat from a baseball from a continent away. No celestial fly balls whizzed by the telescope, but IRAS did discover puffs of cirrus-shaped clouds in interstellar space, as well as a new class of dust-shrouded galaxies, a handful of new comets, and bands of particles out past Mars that may have resulted from countless asteroid collisions. It expanded the list of known infrared sources from a few thousand to a quarter of a million.

When there was time, IRAS scientists could periodically aim the telescope at targets of their choice. Having read Myers's journal articles on low-mass-star formation, Charles Beichman of the Jet Propulsion Laboratory decided it would be worthwhile (unbeknown to Myers) for IRAS's beryllium mirrors to focus on the same clouds that Myers had inspected. While Myers, in addition to his radio work, had struggled for hours on the ground to catch the few bits of infrared radiation reaching the Earth's surface from clouds like Taurus, IRAS was able to scan any one cloud in a matter of seconds.

The results were stunning: In selected clouds and dust globules located within 650 light-years of Earth, so quiescent in their outward appearance, IRAS discovered dozens of young low-mass stars and protostars, objects which are probably less than 100,000 years old. Says Myers, "If you think of this in comparison to the life of a human being, then these stars are in the first days of their lives." The Milky Way, it seems, is a more prolific parent than was once suspected, creating a couple of stars each year. Inside Barnard 5, an unpretentious cloud located in the direction of the Perseus constellation, IRAS detected a bright infrared source that is emitting energy at ten times the rate of our Sun. This conforms nicely with the current idea that stars undergo an extremely luminous phase in the first 100,000 years of their existence.

In fact, astronomers are coming to see that small newborn stars mimic the convulsive actions of their more massive cousins, albeit with much less horsepower. Those "oases of calm" do, at some point, get ruffled. If you think of giant O and B stars coming onto the galactic scene as Cecil B. De Mille spectaculars, then newly made low-mass stars might be considered as putting on the high-school senior play. Take the case of IRS 5, located in a dark cloud called Lynd 1551 about 500 light-years from Earth. This infrared object is probably no more than twice the mass of our Sun; yet, it appears to be shooting off two narrow streams of gas in opposite directions. In radio "pictures," IRS 5 looks very much like a Roman candle burning at both ends. This might be an indication that brisk stellar winds are being channeled outward by a disk of material still surrounding the young sun. This might have been the way the inner regions of our solar system got scoured clean of gaseous debris after its construction. And embedded in IRS 5's jetlike flows, like some kind of celestial glitter, are several glowing blobs of gas called Herbig-Haro objects, each weighing a few Earth masses. These are either chunks of material being shot out of the star, which end up plowing through the interstellar medium at dozens of miles per second, or knots of gas in IRS 5's neighborhood that are getting shocked by the fast stream of matter flowing away from the infant star and moving outward with it.

How did our own Sun emerge from the dusty gas? Was it, like the active young star in Lynd 1551, a big fish in a small celestial pond of primeval material? Maybe not. One piece of evidence arrived in 1969 when a carbonaceous chondrite (a very primitive meteorite distinguished by the presence of carbon and pebblelike chondrules) fell near the village of Pueblito de Allende in northern Mexico. Meteorite experts found that the Allende meteorite once contained a short-lived isotope called aluminum-26 that does not naturally exist on Earth and that the amounts of certain other isotopes were subtly higher than the solar system average.

Some experts conjecture that these uncommon materials were injected into the solar system by a supernova that exploded within our vicinity around the same time that the meteorites were condensing. This may mean that our Sun was not alone in its birth, but was surrounded by quick-burning O and B stars. It would have been a wondrous sight. One can imagine pink streams of ionized hydrogen, activated by the ultraviolet radiation pouring out of the giant stars, painting our primordial nighttime sky from horizon to horizon. The Sun may have been but one of a litter, whose members soon dispersed to parts unknown.

If the Sun was indeed born into a cluster of stars within a giant molecular cloud, it's interesting to speculate whether its sisters and brothers also developed planetary systems. The most favored theories today argue that planets, moons, and asteroids are common by-products of star formation. "Planetary formation should no longer be considered a unique event," says theorist Douglas Lin of the University of California at Santa Cruz, one of many scientists who simulate the process on a computer. "It certainly does not require a catastrophe, as once thought. To the contrary, it looks as if it is a very natural outcome of the star-formation process."

It all begins with a cosmic nudge. Compressed by a shock wave passing through its neighborhood, a roughly spherical clump of gas and dust about a light-year across begins to collapse. Conserving its angular momentum, the condensing cloud spins faster and faster and thus flattens out. The densest region, the protostar in the center, finds itself surrounded by a swirling disk of matter. Individual specks of dust within this disk, grains spewed from old stars and as delicate as a puff of smoke, start colliding, fusing into ever-larger lumps. Over time, the aggregates grow from pebbles and rocks to boulders and larger formations. As these planetary embryos relentlessly orbit the newborn star, they sweep up enough material through accumulation and collisions to form a planet-sized body.

Computer simulations by George Wetherill of the Carnegie Institution of Washington suggest that the last surviving planetesimals wage a ferocious battle, a demolition derby of cosmic proportions, until one planet is victorious in each orbital lane. One of Earth's final collisions may have thrown off the material that eventually coalesced to form the Moon. It is believed that planets situated near the energetic sun, which blasts them with intense radiations, remain rocky; objects farther out in cooler environs, similar to Jupiter and Saturn, are probably able to maintain a thick coat of gas. The entire process takes tens to hundreds of millions of years, a fairly short time when compared to some astronomical events.

Fierce stellar winds might disrupt the development of planets for the most massive stars; for that matter, the fast-burning stars probably expand into red giants or blow up altogether before a full-fledged planetary system has a chance to develop. And the complex gravitational field of a multiple-star system (unlike our Sun, most stars are actually part of a pair or group) could play havoc with a family of planets. But it's still likely that the solar system is not a unique specimen in the cosmos. In a galaxy of 100 billion stars, in a universe of 100 billion galaxies, it seems improbable that our Sun became the sole owner of planetary real estate. Yet, for many years, there was only indirect evidence to support this viewpoint. Many stars had been found to rotate somewhat more slowly than expected. This might mean that the star, while forming, transferred some of its rotational energy to a bevy of satellites. In addition, painstaking measurements of the motions of several stars in our immediate stellar neighborhood have revealed that they deviate ever so slightly from a smooth, straight path, not unlike the wobble of an unbalanced tire. The periodic wiggles suggest that some kind of dark companions, perhaps planets, are tagging along. Anyone watching our Sun traverse the galaxy from a vantage point several light-years away would notice a definite swagger. The Sun sways back and forth almost half a million miles as Jupiter orbits around it.

Estimates on whether extrasolar worlds are circling distant suns improved dramatically right after IRAS probed the heavens. The initial discovery was a fluke. Two IRAS scientists, Harmut H. Aumann of NASA's Jet Propulsion Laboratory in California and Fred Gillett of the Kitt Peak National Observatory, had ordered IRAS to look at Vega, a prominent summer star located 27 light-years from Earth. (Vega is about twice the size of our Sun, and sixty times more luminous.) They intended only to perform a routine calibration of IRAS's instruments, but to their surprise, they found that the fifth-brightest star in the sky emitted more heat radiation than expected. The extra infrared energy, they concluded, came from a disk of material roughly 15 billion miles wide, twice the size of our solar system. IRAS was not capable of resolving the disk, but theory suggests that the particles are pebble-sized or bigger; planet-sized bodies cannot be ruled out. It was one of the first pieces of direct evidence suggesting that nonstellar material orbits another star. Since Vega is less than one billion years old (four times younger than our Sun) this may be material on its way to forming planets, or debris that failed to stick together.

Since that startling discovery, both IRAS and ground-based infrared telescopes have detected the presence of similar protoplanetary rings around a number of other neighboring stars. This new cottage industry took on special significance when University of Arizona astronomer Bradford Smith and Richard Terrile of the Jet Propulsion Laboratory produced the first photograph of a potential solar system outside of our own. Their candidate: an innocuous star in the southern sky called Beta Pictoris, which is located 50 light-years from Earth. Just like Vega, Beta Pictoris was found by IRAS to be emitting a greater amount of infrared radiation than anticipated. Smith and Terrile took a second look with the 100-inch telescope at the Las Campanas Observatory in the Chilean Andes. By blocking out the glare of Beta Pictoris with a special mask and by using a supersensitive electronic camera in place of a photographic plate, the two astronomers were able to discern a thin disk of particles surrounding the nearby, two-solar-mass star. To be exact, their computer-enhanced image showed two streaks of light extending out from either side of the masked star for a distance of some 60 billion miles. It was an image that captivated millions in newspapers around the world.

Astrophysicists conjecture that this faint light emanates from pieces of ice and rock, ranging in size from a thousandth of an inch to a few miles across, not unlike the swarm of cometlike debris surrounding our own solar system. It is intriguing that the innermost region of Beta Pictoris's disk appears to have been swept clean. Such a clearing-out, some astronomers speculate, might be the result of the material having coalesced into a horde of planets, maybe within the last 100 million years. If so, that means that Beta Pictoris offers us the unique opportunity of seeing how our solar system looked more than four billion years ago. At the time of the discovery, Terrile enthusiastically pronounced that "we'll have to realize that, as a civilization, we're not the center of the universe."

The ante was upped in this quest for extrasolar planets when University of Arizona astronomer Donald McCarthy and two colleagues examined the star van Biesbroeck 8 (VB 8), one of Earth's closest neighbors at a distance of 21 light-years, with a special instrument known as a speckle interferometer. In 1983, Naval Observatory astronomers had reported that VB 8 exhibited a definite wobble during its journey through the heavens, the classic sign that the faint star had a small companion gravitationally tugging on it. Normally, such dark companions are lost in the glare of the brighter stars they orbit, but in the case of VB 8 McCarthy was able to detect some infrared radiation emanating from the consort, later dubbed VB 8B. It was the first direct detection of a substellar companion outside our solar system. McCarthy essentially took thousands of snapshots of VB 8 (each snap lasting no longer than a tenth of a second) and built up a signal by processing all these freeze frames with a computer. By using this speckle technique, he was able to reduce the distortions in the stellar image (a star's infamous "twinkle") that are caused by the Earth's turbulent atmosphere.

The infrared emissions gathered by the McCarthy team revealed that VB 8's companion was the size of Jupiter, but ten to sixty times more massive. McCarthy has labeled it a planet, but others prefer to call it a brown dwarf, or near-star, since it is almost weighty enough for internal pressures to trigger the fusion of hydrogen into helium. (About 80 Jupiter masses are needed to ignite a stellar fire.) Though cooler than the coolest star, VB 8B can still generate quite a bit of heat through gravitational pressure. The Arizona team estimates that this gaseous "almost-star" has a surface temperature as hot as a blast furnace, about 2,000 degrees Fahrenheit. By comparison, Jupiter has a surface temperature of −240 degrees.*

All these findings, of course, have a strong emotional pull since they will be helping us answer that ancient question: "Are we alone?" The chances that we're a solitary outpost in the boondocks of this galaxy look slimmer and slimmer with each new discovery. "I view planets as sort of cosmic petri dishes," says NASA astrophysicist David Black. "If planets are plentiful, then you have a stronger feeling that intelligent life can exist within a reasonable distance of our solar system. But finding one isolated low-mass object is not enough. We're never going to understand the origin of our own solar system until we can compare it with many other systems, stars with two or more dark companions. For example, are all planetary systems structured alike, gas giants at the edge with a few insignificant pebbles like Earth rotating around inside?"

Direct imaging of more Earth- or Jupiter-like bodies outside our solar system could come with future generations of infrared telescopes launched into space in the 1990s and the 21st century. A "Jupiter" might appear as a small infrared dot that slowly moves around the star, not unlike Galileo's first view of Jupiter's moons almost four centuries ago.

*The existence of VB 8B became questionable when other astronomers failed to observe it. Additional brown-dwarf candidates, though, have since been sighted.

It could be the bait that finally lures humankind itself out of the solar system. Muses veteran planet-hunter George Gatewood of the University of Pittsburgh's Allegheny Observatory: "Man has never had a carrot dangle in front of him that he didn't go after sooner or later. If Man finds an Earth-sized planet circling a nearby star, he's eventually going to probe it," just as humanity was enticed for untold generations by a satellite called the Moon.

Just a century ago, it was not really clear to astronomers that stars continue to be conceived and, once gestated, go on to experience different stages of development, akin to a caterpillar. From the viewpoint of a nineteenth-century observer, all the stars in the sky could well have lighted up simultaneously at the beginning of the universe. Even as late as the 1940s, when the world's top astronomers gathered at a symposium to discuss what they knew about star formation, theories were limited, supported only by the meager evidence gathered by optical telescopes.

But astronomy's quest to inventory the gaseous ingredients floating in the vast ocean of space between the stars resulted in a pleasing payoff. "Thanks to the development of new telescopes and equipment for the detection of radiation at radio and infrared wavelengths, astronomers are finally able to explore the dark and dusty clouds where stars are born," says Lada. "And by deciphering the mysteries of star formation, we are also learning about the creation of planets. Eventually, we may understand not only how the solar system formed, but ultimately how life itself came into being."

But stars cannot live forever. Sooner or later, they must exhaust their fuel supplies. Determining what happens to a star at the end of its life cycle has been a major theoretical and observational undertaking for astronomers throughout most of the twentieth century. The old guard, trained to expect a less complicated universe, were somewhat surprised to find that stars can die in a variety of ways, depending on the star's mass and its immediate environment. The biggest stars blow up in titanic explosions; the smallest simply cool, quite slowly, into dark planet-sized cinders. Over the last few decades, arsenals of equipment—radio and optical telescopes on the ground, x-ray and ultraviolet detectors in space—have been aimed at our galaxy's senior citizens, and the results have compelled astronomers to rewrite the story on the fate of stars.

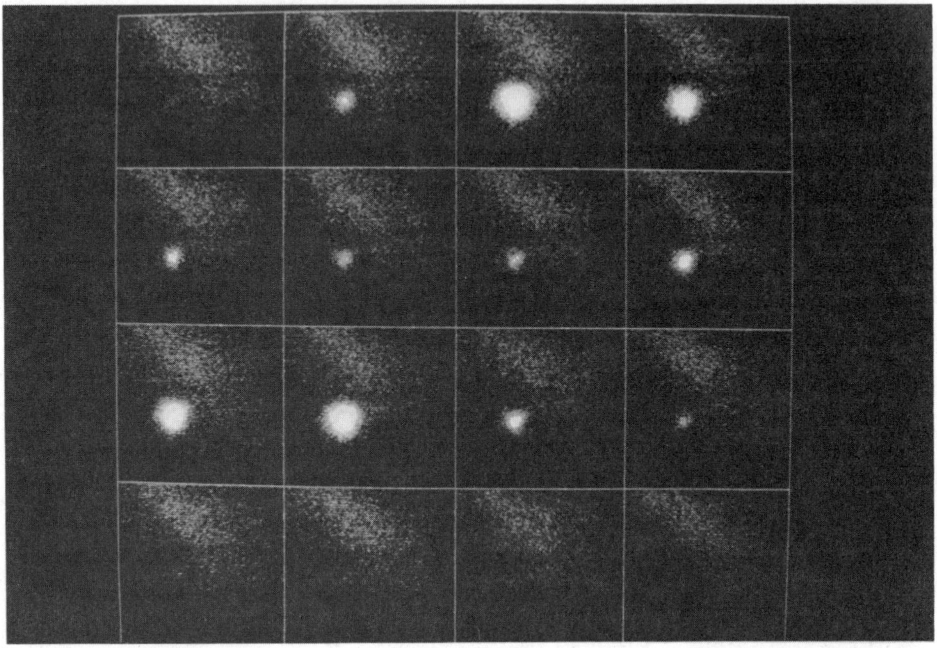

The Crab pulsar, a relatively young neutron star embedded in the Crab nebula, blinks on and off like a celestial lighthouse as it spins. The main pulse peaks in the third frame (from left to right); a secondary pulse, half as strong, reaches maximum brightness in the tenth. The entire sequence (one complete rotation of the ten-mile-wide star) takes one-thirtieth of a second. *Courtesy of F. R. Harnden, Jr., and F. D. Seward, Harvard-Smithsonian Center for Astrophysics, Cambridge, Massachusetts.*

2

A TWILIGHT'S LAST GLEAMING

Do not go gentle into that good night,
Old age should burn and rave at close of day;
Rage, rage against the dying of the light.

—Dylan Thomas, *Do Not Go Gentle into That Good Night*

It was noon on a crisp wintry day in the Maryland suburbs of Washington, D.C., when the radio antenna, one of many situated on the campus-like grounds of NASA's Goddard Space Flight Center, transmitted a string of commands to the International Ultraviolet Explorer (IUE), a space telescope hovering some 20,000 miles above the Atlantic Ocean. The astronomer scheduled for that afternoon's eight-hour observing run, Harry Shipman, of the University of Delaware, had just asked the telescope operator to aim the IUE's sensitive spectrographs toward a faint stellar object labeled GD 356 in astronomical catalogs. With a few expert keystrokes on a computer console set in the middle of the cramped, equipment-filled control room, the radio signal was sent out, eventually activating a flywheel mounted on the high-flying telescope. As the flywheel spun one way, the 1,800-pound satellite responded by rotating ever so slowly, in perfect illustration of Newtonian mechanics, in the opposite direction in search of its next target.

Launched in 1978 as a cooperative venture between Great Britain, the European Space Agency, and the United States, the IUE is accessible to large numbers of astronomers and able, because it bypasses our absorbent atmosphere, to collect the ultraviolet radiations from a wide range of celestial events, from auroral displays in the polar regions of the planet Jupiter to stellar explosions in far-off galaxies.

On this particular day, Shipman was using the IUE as a kind of time machine: GD 356 is an extremely tiny star, known as a white dwarf, whose canescent and withered appearance foretells what our own Sun, the nearest star, will look like in the year 7,000,000,000 A.D. "The total mass of GD 356 is not much different from that of our Sun, but its girth is a hundred times smaller, having collapsed after expending all of its useful fuel," explains Shipman, who has devoted most of his academic career to researching these aged stars. For GD 356, this is the final curtain, the twilight period in a life that probably began with great verve in some transient molecular cloud. For every birth, there is an inevitable death.

Today, in the prime of its life, the Sun has a diameter of nearly one million miles. That's four times the distance from here to the Moon. "But once the nuclear fires of youth vanish from its interior," says Shipman, "the Sun will eventually shrink down to an ultradense ball about the size of the Earth." It will become a close cousin to GD 356, which contains roughly two billion billion billion tons of matter in its planet-sized volume. Just a few drops of this white-dwarf stuff weigh 2,000 pounds; two cups would outweigh all the cars parked at a shopping mall during a holiday sale. Most stars inhabiting the Milky Way, those staunch bourgeois with about the same mass as the Sun, will ultimately arrive at this fantastic compression. Caltech astronomer Jesse Greenstein, Shipman's mentor and a pioneer in the study of white dwarfs, has said that 100 billion years from now these lilliputian stars will inherit the skies by simple default. Astronomers estimate that ten billion stars scattered throughout the Milky Way, one-tenth of the galaxy's stellar population, have already experienced this tremendous gravitational squeeze. An exact count is hard to come by, since white dwarfs are so faint that not one is visible to the naked eye. By examining GD 356 with the IUE, Shipman was witnessing a portent of things to come. The small white star is no more than a dense ash, the end point in a long and complex stellar metamorphosis.

Stars more massive take a radically different route to stellar demise. They "do not go gentle into that good night" but instead tear themselves apart in a spectacular explosion, leaving behind supercompressed remnants, each no bigger than a city, that at times signal their presence with regular pulses of radio or x-ray energy. Not until the middle of this century were such varied pathways in stellar evolution fully comprehended, for before astronomers could decide how stars die they had to figure out how stars burn.

Pity the poor nineteenth-century astronomer who, before thermonuclear fusion was fathomed, attempted to explain why the Sun shone. An understanding of chemical energy, such as the burning of fossil fuels, didn't help at all; if the Sun were one huge lump of coal, emitting the radiations that currently warm the Earth's surface, it would burn up completely in 10,000 years. To get around this shortcoming, Victorian scientists imagined a variety of (in retrospect) outlandish mechanisms. The English physicist Lord Kelvin once proposed that a persistent rain of meteorites, kindled by their fall into the Sun, served as a source of life-giving radiations. Later, he and Hermann von Helmholtz would suggest that gravitational collapse could keep the Sun hot and glowing. They surmised that as gravity pulls the stellar gases inward, heating them up, some of the energy is radiated outward. The two physicists calculated that the Sun merely has to shrink some 70 feet a year to account for its present luminosity, a rate of contraction that would hardly be noticed over the span of recorded history. In this way, Kelvin and Helmholtz deduced, the Sun could shine for a few dozen million years.

This seemingly reasonable conclusion became untenable, however, once evidence from the fossil and geologic records hinted that the Earth was more than one *billion* years old.* Having an Earth older than the

*Today, the ages of meteorites—those rocky leftovers from the birth of our solar system that sometimes plunge through the Earth's atmosphere with a fiery bravado —tell us that the Sun and its attendant planets formed about four-and-a-half billion years ago.

Sun was a scientific embarrassment, to say the least. Spurred by the emergence of such revolutionary new fields as quantum mechanics and particle physics, Sir Arthur Eddington of Great Britain, one of the most distinguished astrophysicists of his day, wrote in the 1920s that it was "necessary to assume that sub-atomic energy of some kind is liberated within the star, so as to replenish the store of radiant energy." His remarkable insight, however, was rebuffed by a number of critics, who argued that the Sun's interior was simply not hot enough to spark nuclear fusion. Eddington readily responded, with characteristic aplomb, that they should "go and find a hotter place!"

There was no need. By 1939, Hans Bethe in the United States and Carl von Weizsäcker in Germany had independently shown, in a series of elegant equations, how stars can be powered by the welding of four atoms of hydrogen into the heavier element helium, whose nucleus contains two protons and two neutrons. Today, physicists are attempting to mimic a similar reaction on Earth, albeit on a much smaller scale, either by confining a hot, electrically charged gas called a plasma within an intense magnetic field or by zapping pellets of heavy water with a high-powered laser beam. Bethe and von Weizsäcker were the first in a long line of researchers who revealed that stars are no less than nuclear chemical factories, which daily manufacture a whole assortment of light and heavy elements.

In fusing hydrogen into helium, stars derive their energy in much the same way that a hydrogen bomb does. Normally, hydrogen nuclei, or protons, are repelled by one another, since they are all positively charged. But deep in the heart of a star, where temperatures reach over ten million degrees and densities are twelve times that of lead,* protons can gain enough energy to collide with a force that sometimes overcomes that electromagnetic repulsion. This enables some of the protons, following an involved chain of reactions, to stick together with a strong nuclear glue. In this manner, our own Sun converts 594 million tons of hydrogen into 590 million tons of helium with each tick of the clock. The four million tons lost along the way is the mass which is trans-

*By racing around at breakneck speeds, though, electrons and nuclei in a stellar core continue to behave as a gas. Halfway out from the center, the solar density is about equal to water; at the surface it is one ten-thousandth that of air.

formed into pure energy according to Albert Einstein's famous equation $E = mc^2$ (energy equals mass multiplied by the square of the speed of light). This four-million-ton conversion each and every second is what bathes the entire solar system in heat and light. The Sun has been doing this for nearly five billion years and has enough hydrogen fuel in its core to continue for five billion more.

Over these eons, the Sun performs the universe's most amazing balancing act, as do all stars. While the ever-present force of gravity pulls inward, trying to squeeze the stellar material tighter and tighter, the pressure of the hot gases and particles bouncing around inside the Sun, energized by the nuclear reactions deep in the core, acts as a counterbalance. As a result, the Sun neither wafts away nor shrinks into oblivion. The two opposing forces in this battle are at a stalemate.

But all good equilibria must come to an end. Five billion years from now, when the bulk of the hydrogen in the Sun's core is converted to helium, the central furnace will flame out. Nuclear "burning" will then take place in a shell of hydrogen surrounding the inert helium core—and at a furious rate that makes the Sun grow brighter and brighter. At this point, the Sun will experience one of the most dramatic transformations that can take place during the life of a star: conversion into a cool, bloated star known as a red giant. It is the astronomical equivalent of a caterpillar's metamorphosis into a butterfly. If you could magically transport yourself to a ringside seat far above the solar system during this far-off epoch, you would observe something like this:

Like the fascinating Siamese creature in the Doctor Dolittle stories, the aging Sun will take on a sort of "pushmi-pullyu" quality. Because the nuclear fire in the very center of the star has been extinguished, the dormant helium core, which is about the size of a giant planet, will start to contract. But in doing this, the core gives up gravitational energy that will, conversely, *push* the outer envelope of the Sun farther and farther outward. This formerly unremarkable star will swell to gigantic proportions, changing its predominant hue from yellow to red as temperatures decrease in the greatly expanded surface layers. Needless to say, the inner planets, Mercury, Venus, and Earth, will sustain a change in their surroundings that could rival Dante's Inferno. After the Earth's miles-deep oceans have boiled away, you will see our planet's longest mountain range: the craggy Mid-Atlantic Ridge that runs like a spinal column from Iceland down past the tip of South America.

During its development as a red giant, the Sun will undergo several episodes of expansion and contraction over a period of hundreds of millions of years, as adjustments are made within its nuclear-fired furnace; it might even pulsate at some point as a variable star. But once it reaches maximum extension, the bloated globe could very well fill Earth's orbit, swallowing the inner planets. For some tens of thousands of years, the rubylike orb will outshine most stars in this outlying sector of the Milky Way, nearly equaling the luminous splendor of the notable red giants shimmering in the present-day sky, such as Betelgeuse (the prominent star that marks Orion's right shoulder), ruddy Aldebaran (the gleam in the eye of Taurus), and mighty Antares (Scorpio's brightest star, whose name signifies "rival of Mars").

During a star's red-giant phase, new fusion reactions come into play. Helium is no longer quite so stable. With the hydrogen-burning shell heaping more and more helium upon the core, the latent center becomes enormously compressed, and likewise heated. Once the temperature reaches 100 million degrees inside the stellar core, the helium finally ignites and fuses into carbon and oxygen. "This fusion process can, in some cases, go on and on," says Roger Chevalier, of the University of Virginia. "The carbon core gets surrounded by a helium-burning shell, with a hydrogen-burning shell farther out. In this way, the center of a red giant starts developing a series of layers, like an onion skin."

Our bantamweight Sun will stop thermonuclear fusion at the carbon-core stage. But in stars more massive, the carbon and oxygen atoms go on to fuse into neon and magnesium. These, in turn, can serve as the raw materials in the construction of even heavier elements, such as silicon, sulfur, argon, and calcium, each chemical group burning in successive concentric shells. If a star is massive enough, its core will continue fusing elements until iron is formed, which is the end of the line. The fusion of iron nuclei doesn't generate any power; to the contrary, their union requires an input of energy. (Elements heftier than iron do get made during a star's lifetime, lightly sprinkled throughout the star's core like some added spice, but these are side reactions which occur much more rarely.)

This picture of a star's developing anatomy was drawn as scientists

—most notably, Margaret and Geoffrey Burbidge, William Fowler, and Fred Hoyle—applied the equations of nuclear physics to a stellar environment. The group's 1957 paper on the synthesis of elements inside a star is such a classic that astronomers refer to it, in scientific shorthand, as "B^2FH." Since its publication, observers, such as Chevalier, have gone on to confirm many of the paper's conclusions by telescopically sifting through the remnants of exploded stars to deduce a star's inner composition.

For instance, in analyzing the light emanating from debris shot out of a stellar explosion known as Cassiopeia A, Chevalier saw that the stellar scraps harbored little hydrogen, but did contain lots of oxygen, sulfur, argon, and calcium. "It appears that this material has been through oxygen burning to form the silicon-group elements, one of the advanced stages of nucleosynthesis," he says. "Not only do observations of this remnant show how heavy elements are ejected into the interstellar medium but also detailed abundance studies of it make possible a comparison of nucleosynthesis theories with observation. We have a rare opportunity to see into the core of a massive star."

Astronomers who point their telescopes at red giants these days are finding that the corpulent stars are the galaxy's worst litterbugs. The intense radiations that originate in a red giant's blazing heart move to the surface and generate brisk stellar winds that can blow away large portions of the star's diaphanous envelope at speeds of dozens of miles per second. For example, using the same speckle interferometry techniques that enabled Donald McCarthy to detect the substellar companion VB 8B, University of Arizona astronomers have photographed Betelgeuse in the act of shedding its outer garments in successive layers. This cosmic "striptease" is one major way in which stars seed the interstellar medium with element-rich material for the next generation of stars and planets. With thousands of red giants undressing at any one time, enough material is cycled back into the galaxy each year to form several new suns. The amount of matter whisked off the surface of any one red giant each day and ejected into space can be up to ten billion times greater than the mass of particles our own Sun loses daily through the action of its solar wind. A waning red giant, in its last gasp, might even throw off its remaining outerwear during a few eruptions.

The shells of material that have been wafting off the star for hundreds of thousands of years balloon into space. For the red giant of relatively small mass, in the range of the Sun, this process eventually exposes the giant's hot carbon core, which will shrink, cool, and then die as a remnant dwarf star. White dwarfs, like Shipman's GD 356, are merely the compressed innards of former red giants. Astronomers can tell exactly when a red giant's veil is lifted; a glowing sign points the way. The white dwarf's energetic radiations cause the ever-expanding bubble of gas—the red giant's discarded raiment—to glow softly as a roundish "planetary" nebula. Planetary nebulae got their name when early astronomers first sighted the rings through small, crude telescopes. They thought the structures resembled the disks of planets. These fluorescent signs stay on for a few tens of thousands of years, until the gas disperses. It is the galaxy's way of announcing, "The Red Giant is dead. Long live the White Dwarf!"

Recognition of the white dwarf as a distinct stellar object came only in the first decades of this century, as astronomers were attempting to decipher the strange spectral signatures of several anomalous stars. In astronomy it is often the celestial oddball that enables observers to reveal a more general truth. Particularly troubling at this time was a very faint star, slightly more massive than our Sun, which circles about the brightest star in the nighttime sky, Sirius, once every fifty years. In examining the spectral features of Sirius's dim companion from atop Mount Wilson in California, Walter Adams discovered, to his amazement, that the orb was burning with a white-hot fury.

Adams knew that a star appreciably hotter than our Sun should be many times brighter; yet, he could see that if Sirius's companion* changed places with our Sun it would appear 400 times *less* luminous. There was only one way this paradox could be resolved: The total surface area emitting those intense radiations had to be extremely small. A great mass, about equal to that of our Sun, was squeezed into a volume as compact as a planet. (A tiny hot star simply puts out less light than a bigger star at the same temperature.) Eddington described his bemusement at the news thus: "The message of the Companion of Sirius, when

*Because of its location in the constellation Canis Major, Sirius has long been known as the Dog Star. Inevitably, Sirius's companion was dubbed the "Pup."

it was decoded, ran: 'I am composed of material 3,000 times denser than anything you have come across; a ton of my material would be a little nugget that you could put in a matchbox.' What reply can one make to such a message? The reply that most of us made in 1914 was—'Shut up. Don't talk nonsense.' "

By 1926, the stellar hieroglyphics were clearly translated, when the English physicist Ralph Howard Fowler figured out that Sirius's companion was a completely new state of matter, an ultradense assemblage of particles impossible to construct on Earth. Since the thermonuclear engine inside a white dwarf has been turned off, having run out of fusible fuel, the star contracts, and consequently heats up. In the lengthy war between gravity and radiation pressure, gravity wins the final battle. Temperatures inside the dwarf became so extreme that its atomic nuclei are stripped of electrons, enabling the white-dwarf material to be crushed to unearthly densities. The compression continues until all the electrons and nuclei in the star, like droves of tiny billiard balls, are packed into the smallest volume possible. A pressure exerted by the electrons resists further compaction. The radiation that a white dwarf continues to emit is merely the energy left over from its more fiery past.

The IUE space telescope has been particularly adept at examining white dwarfs, since these intensely hot stars emit large amounts of ultraviolet radiation in their "youth." But, like the cooling embers that they are, white dwarfs will ever so slowly turn yellow, then orange, then red. After billions of years, they will become mere black, crystalline cinders resting, apparently in peace, in the graveyard of space. If the core of a white dwarf is composed largely of supercompressed carbon, that old refrain about a twinkling little star being "like a diamond in the sky" may not be too far from the truth. As far as astronomers know, such will be the fate of the Sun. Such will be the denouement for most of the suns in the heavens. Each will quietly exit as a black dwarf—that is, unless they are accompanied by a particularly close companion.

If all stars were like our "bachelor" Sun—carefree and single in its journey through the galaxy—that would be the end of the white-dwarf story. But more than half the stars in the Milky Way are actually members of multiple systems, where two or more stars orbit one another the way the Moon circles the Earth. Under these circumstances, an elderly white dwarf has a chance at experiencing a more interesting fate. Before winking out of existence, it might kick up its heels a bit in a spectacular manner.

Kentaro Osada, a Japanese amateur astronomer, observed some of this dwarfish carousing on the night of August 29, 1975. As twilight settled over Japan, Osada noticed that Cygnus the Swan, a prominent constellation in the summertime sky, had a star in its tail that wasn't there the previous night. The newcomer's luminosity soon rivaled that of Deneb, the brightest star in Cygnus. Before the Earth could complete half a rotation, hundreds of other amateur and professional astronomers wired or phoned news of the startling appearance to the Central Bureau for Astronomical Telegrams in Cambridge, Massachusetts, the official clearinghouse for new celestial sightings. "I remember the night well," says astronomer Brian Marsden, head of the bureau. "It completely spoiled my Labor Day weekend." What those hundreds of astronomers were witnessing was a nova, a name given to the event by ancient astronomers in their mistaken belief that the heavens had created a brand-new star. In actuality, a classical nova is the temporary rejuvenation of a very old member of the stellar corps.

Astronomers surmise that novae, such as Nova Cygni 1975, originate in systems where a white dwarf is paired, more closely than usual, with another star. The two stellar objects can be separated by less than 500,000 miles, a distance smaller than the width of our Sun—they're practically hugging! "Obviously, something dramatic has to happen to bring these two stars so close," points out nova expert Richard Wade, of the University of Arizona.

In a tightly packed swarm of stars, such as a globular cluster, where members are separated by light-minutes instead of light-years, the two stars may have gravitationally coupled as they passed very near one another. Astronomers have come to believe that in the more open spaces of the galaxy, it takes a bit of stellar cannibalism within a binary-star system. "During the white dwarf's former life as a red giant," explains Wade, "its bloated envelope may have swallowed up the other star, causing the less-massive companion to spiral in. The swallowed star is not greatly affected, since a red giant's outer layers are very tenuous, but the event does set the stage for future developments."

It creates a system whereby a white dwarf and a companion star can be intimate neighbors—so intimate that the dwarf's intensely strong gravitational field begins to pull gas, predominantly hydrogen, away from the companion's outer atmosphere. Like water circling a drain, the river of gas streaming off the star forms a swirling disk of matter around the white dwarf, akin to Saturn's rings only bigger and fluffier. Astronomers call it an accretion disk. Over time, some of the disk material reaches the surface, gradually wrapping the entire dwarf in a thin blanket of hydrogen. Compressed and heated by the terrifically high gravity, the hydrogen layer eventually ignites, engulfing the dwarf in a monstrous thermonuclear conflagration. A nova is born.

Astronomers estimate that about thirty white dwarfs sprinkled throughout our galaxy blow off a little "steam" each year in this manner, but most are hidden by interstellar dust. Three or four of the novae do get spotted, owing largely to the efforts of amateur astronomers, like Osada, who scan the heavens nightly. In a typical nova, the white dwarf brightens 100,000 times over; it releases, in that one explosive moment, more energy than our Sun emits over 100,000 years. Nova Cygni, or V1500 Cygni as it is now officially labeled, was even more impressive. Its luminosity increased by a factor of 100,000,000. If such a nova were to occur in the vicinity of Alpha and Proxima Centauri, the two stars nearest the Sun, the brilliant point of light would outshine the Moon for many days before fading back into celestial obscurity. Though a nova is a rather violent outburst, the system remains intact and can repeat its explosive behavior. Nova Cygni might reappear in 10,000 or more years, after the system's thief, the white dwarf, has the opportunity to steal away another layer of fusible hydrogen gas from its companion.

As soon as the riddle of the white dwarf was unraveled in the 1920s, astronomers smugly assumed that all stars—red stars, blue stars, tiny ones, fat ones—inevitably reached this final stage. But an insatiably curious student from India, working on his Ph.D. at Cambridge University in England during the early 1930s, radically altered this view, and in face of great opposition from some of the most illustrious scientists of the era. Subrahmanyan Chandrasekhar applied Einstein's Theory of Special Relativity to the equations describing the structure of a white dwarf and realized, from his straightforward exercise, that stellar remnants more weighty than a certain mass (now considered to be approximately 1.4 solar masses) would never settle down as stable white-dwarf stars in their old age, but rather would continue to contract. Later, after the dust had settled from the brouhaha generated by this controversial calculation, the 1.4-solar-mass dividing line between white dwarfdom and further collapse to some stellar netherworld came to be known as Chandrasekhar's limit. It was the first step in a long career devoted to the study of stellar evolution and dynamics, one that would lead the University of Chicago theorist to the Nobel Prize in 1983.

But in the early thirties, Chandrasekhar was still the neophyte within the hallowed world of Cambridge astrophysics. In 1932, at the age of twenty-two, the young man from India wrote that "great progress in the analysis of stellar structure is not possible before we can answer the following fundamental question: Given an enclosure containing electrons and atomic nuclei, what happens if we go on compressing the material indefinitely?" This was a very disturbing question for astrophysicists to consider; many hoped it would just go away. Eddington, with whom Chandrasekhar was acquainted at Cambridge, was the most vocal opponent: "The star apparently has to go on radiating and radiating and contracting and contracting until, I suppose, it gets down to a few kilometers' radius when gravity becomes strong enough to hold the radiation and the star can at least find peace," he stated at a 1935 meeting of the Royal Astronomical Society. "Various accidents may intervene to save a star, but I want more protection than that. I think there should be a law of nature to prevent a star from behaving in this absurd way."

At the time, Eddington did not realize that his glib remark concerning miles-wide stars, meant only to serve as a preposterous fantasy, would turn out to be a telling prophecy. In his refusal to accept Chandrasekhar's conclusions, this renowned astrophysicist missed the chance of recognizing one of the universe's rarest and most fascinating denizens: the neutron star.

Contemplating the physical characteristics of a neutron star is a frustrating endeavor, mainly because the description defies common sense. You get the feeling, after a while, that you have entered the proverbial "twilight zone." White dwarfs are bad enough. A teaspoon of white-dwarf matter, you may recall, weighs more than a car. But as astrophysicist Jonathan Grindlay from the Harvard-Smithsonian Center for Astrophysics points out, "The density of a neutron star is roughly what would be achieved by packing *all* the automobiles in the world into a thimble!"

Chandrasekhar had asked what would happen to a star if it were compressed beyond white-dwarf densities. The answer: The negatively charged electrons and positively charged protons would ultimately merge to form a compact ball of neutral particles called neutrons (hence the tag, neutron star). The fact that matter can be compressed to such extreme densities goes to show that atoms are mostly empty space: small, dense nuclei surrounded by a tenuous and widespread cloud of electrons. If a nucleus were blown up to the size of a golfball, the electrons would be scattered over miles. A neutron star, the mass of about one-and-a-half suns squashed into a sphere around a dozen miles in diameter, merely eliminates all that excess room. Its atoms are crushed completely. In some sense, a neutron star can be considered to be one huge atomic nucleus that happens to contain a billion trillion trillion trillion trillion neutrons.

On a star of such incredible density, gravity has a strength of mythical proportions. To fully escape Earth's gravitational grasp, a rocket has to speed off into outer space at a velocity of seven miles per second. To break loose from the clutches of a neutron star, that same rocket would have to travel at speeds of 100,000 miles per second, more than half the speed of light. Neutron-star "mountains," for that matter, can tower no higher than a fraction of an inch. Upon hearing these bizarre statistics, one can develop a sympathy for Eddington's position that a star should not behave "in this absurd way."

The concept of a neutron star was actually introduced in the early 1930s when Soviet physicist Lev Landau suggested that the compressed cores of massive stars might harbor "neutronic" matter. Neutrons were a hot item at the time, having been recently discovered by experimentalists. Astronomers Walter Baade and Fritz Zwicky picked up on the idea and proposed that under the most extreme conditions—during the explosion of a star, to be exact—ordinary stars would transform themselves into naked spheres of neutrons. But their proposal was considered wildly speculative, and only a handful of physicists, including J. Robert Oppenheimer, who went on to become the father of the atom bomb, even bothered to ponder the construction of a neutron star. For three decades, neutron stars remained mere theoretical playthings—until some extraterrestrial "beeps" turned theory into reality.

In radio-astronomy circles, the unflagging persistence of Jocelyn Bell (now Burnell) in tracking down some "bits of scruff" on miles of chart-recorder paper has become legendary, almost as famous a tale as Isaac Newton's observing the falling apple while contemplating gravity. The year was 1967. A small platoon of students and technicians had just finished constructing, near Cambridge University, a sprawling radio telescope: more than two thousand dipole antennas, lined up like rows of corn, that were connected by dozens of miles of wire. The entire complex covered an area the size of fifty-seven tennis courts. Bell, a Cambridge graduate student and native of Ireland, was one of the laborers. "I like to say that I got my thesis with sledgehammering," she jokes these days.

The telescope, designed by Cambridge radio astronomer Antony Hewish to search for extremely distant and energetic galaxies called quasars, had a prodigious output as it passively scanned the sky overhead, and it was Bell's job to analyze the veritable river of data for her doctoral thesis. At a conference commemorating the fiftieth anniversary of Jansky's discovery of cosmic radio waves, Bell recalled the moment when she realized that some of the squiggles recorded on the reams of strip chart paper didn't look quite right:

We had a hundred feet of chart paper every day, seven days a week, and I operated it for six months, which meant that I was personally responsible for quite a few miles of chart recording.

It was four hundred feet of chart paper before you got back to the same bit of sky, and I thought—having had all these marvelous lectures as a kid about the scientific method—that this was the ideal way to do science. With that quantity of data, no way are you going to remember what happened four hundred feet ago. You're going to come to each patch of sky absolutely fresh, and record it in a totally unbiased way. But actually, one underestimates the human brain. On a quarter inch of those four hundred feet, there was a little bit of what I call "scruff," which didn't look exactly like [man-made] interference and didn't look exactly like [quasar] scintillation. . . . After a while I began to remember that I had seen some of this unclassifiable scruff before, and what's more, I had seen it from the same patch of sky.

The 81.5-megahertz radio signal was emanating from a spot midway between the stars Vega and Altair.

Bell discussed the finding with Hewish, and they decided to observe the peculiar signal with a higher-speed chart recording. The "scruff," on the other hand, had other ideas. "Every day for the best part of November, I went out to the observatory and switched on the fast chart recorder and got lovely recordings of receiver noise, but no sign of the scruff," remembers Bell. "Tony Hewish was getting a bit peeved."

Fortunately, at the end of the month, the signal popped back up, and Bell was clearly surprised by what the more detailed recording revealed. She saw that the signal was actually a precise succession of pulses spaced 1.3 seconds apart. The astonishing regularity in the timing of the pulses would make a Swiss watchmaker proud. The unprecedented, clock-like beeps caused Hewish and his group to label the source LGM for "Little Green Men." This was done only half in jest. At one point, some consideration was given to the possibility that the regular pulsations were coming from a beacon set up by an extraterrestrial civilization. Bell was a bit annoyed at the thought: "I was now two-and-a-half years through a three-year studentship and here was some silly lot of Little Green Men using *my* telescope and *my* frequency to signal the planet Earth."

Right before Christmas, though, Bell ferreted out, from the yards upon yards of strip chart that were ever spewing from the telescope, the telltale markings of a second suspicious source. A late-night trip out to the telescope clinched it. "It was another string of pulses, 1.19 seconds' period this time," she says. For Bell, finding LGM 2 was a great relief: "It was highly unlikely that two lots of Little Green Men could choose the same unusual frequency and unlikely technique to signal to the same inconspicuous planet Earth!" By the beginning of 1968, two more were found. As soon as the phenomenon was announced to the public, a British journalist dubbed the freakish sources *pulsars.*

Finding pulsars became the astronomical rage. And while observers eagerly searched the heavens, theorists hurried to their desks to generate a blizzard of journal articles surmising how a celestial object (or objects) might emit those regular beeps. Within a year of Bell's discovery, it was generally agreed that rapidly rotating neutron stars fit the bill. The idea that the universe could construct compact stars, whose widths would barely stretch the length of Manhattan, was no longer wild speculation.

Such will be the fate for most aging giant stars, of masses from eight to more than thirty times our Sun's, whose collapsing interiors pass beyond Chandrasekhar's limit on the weighing scale.

Theories concerning radio pulsars are constantly being refined and challenged, but two properties are crucial in any explanation of the pulses: spin and magnetism. As a star collapses, compressing the stellar material ever tighter, it must spin faster and faster, much the way a figure skater turns into a whirling blur as she brings her arms in to her body during a spin. This is simply the law of conservation of angular momentum at work. Concomitantly, the star's magnetic field becomes enormously compressed, and hence intensified. The magnetic field at the Earth's surface has the strength of a toy magnet (just strong enough to swing a compass needle northward). A neutron star's magnetic field, on the other hand, is more than a trillion times stronger.

Because of a pulsar's rapid spin, the highly magnetized body becomes no less than an electrical generator—one of gigantic proportions. The trillions of volts produced by this stellar power plant enable swarms of charged particles to pull away from the surface of the star and race into space. Guided by magnetic field lines, most of the particles are channeled into two narrow beams, which shoot out in opposite directions from the star's north and south magnetic poles. Accelerated to near-light velocities, these two sprays of escaping particles emit a continuous scream of radio energy. We detect the beamed radio emissions as pulses because the spinning neutron star acts like a lighthouse. As on Earth, the pulsar's polar magnetic-field lines are not exactly parallel to its spin axis. So, as the compact body rotates, the radio beams, tilted from the axis of rotation, periodically sweep across earthbound telescopes, much the way a lighthouse beam regularly skims across a coastline. With each rotation we observe, depending on the pulsar's alignment with Earth, either one or two "blips" of radio energy.

Presently, more than 350 radio pulsars have been found throughout our galaxy, out of a probable population of a few hundred thousand. The slowest ones manage to emit a pulse about once every four seconds; others chatter on at fantastic rates. In 1983, Donald Backer of the University of California at Berkeley surprised the entire astronomical community by detecting a pulsar in the constellation Vulpecula that emits a burst of radio energy every 0.001558 second. That means this cosmic lighthouse is spinning at the dizzying rate of 642 revolutions per second! This is fairly close to the limit at which a neutron star can spin without being ripped apart by centrifugal force. A fast spin is usually the sign of a youthful pulsar, but this "millisecond" pulsar could be an exception. Some observers believe it's actually an old neutron star that got spun up as it siphoned material away from a now defunct companion.

These whirling dervishes are more than celestial curiosities. "The bursts of radio energy sent out by pulsars serve as powerful probes of the interstellar medium," pulsar hunter David Helfand of Columbia University once noted. "And as more are discovered, they become useful markers with which to delineate the structure of our galaxy." Sometimes, a pulsar can develop a glitch in its signal; the pulsing will speed up ever so slightly. Astronomers believe this is caused by a "starquake," a sudden cracking or slippage of the star's solid outer crust as it interacts with a faster-spinning superfluid core. Thus, the ticking of the pulsar clock can serve as a sensitive seismometer that allows astronomers to plumb the interior of a neutron star, as though they were cosmic geologists.

You can think of a pulsar as a stellar tombstone, which marks the spot where a giant star, too heavy to die quietly as a white dwarf, long ago tore itself apart in a brilliant supernova explosion. Such a calamitous event makes weaker outbursts, like Nova Cygni, appear decidedly wimpish. For many days, even weeks, the white-hot debris of a supernova can outshine an entire galaxy's worth of stars. The prefix *super* is not a frivolous overstatement, for a supernova is not an explosion *on* a star, as with a simple nova; it is the very explosion *of* a star. Zwicky and Baade had imagined, as far back as the 1930s, that neutron stars might be forged in the fiery blast of a supernova, but this suspicion wasn't fully confirmed, even after Bell's discovery, until the comforting beat of a pulsar—30 pulses per second, to be exact—was detected in the very heart of astronomy's most famous supernova remnant, the Crab nebula.

In terms of absolute energy, it was the most stupendous Fourth of July fireworks the firmament has ever displayed, and this was 722 years *before* John Hancock put his famous signature to the Declaration of

Independence. In the wee hours of the morning of July 4, 1054, the Chinese court astrologer Yang Wei-Te noted that a "guest star" had taken up residence in the eastern sky. So brilliant was its light that the new arrival was visible in daylight for more than three weeks. The stellar visitor, which according to Yang burned with "an iridescent yellow color," didn't disappear completely until 650 days later. A memento of that visit, however, still remains. When modern-day astronomers point their instruments in the direction of the star's former residence in the constellation Taurus the Bull, they observe a tangled web of orangy-red filaments, six light-years wide, that weaves in and out of a cloud of bluish-white light.

The eerie white glow is now known to be caused by hordes of energetic electrons spiraling about in magnetic fields, all pumped up by the spinning pulsar embedded in the bowels of the billowing cloud. The reddish streamers, the remnant debris of the exploded star, are proof of the power of a supernova; nine centuries after the blast, the filaments still move outward at hundreds of miles a second. Using his then revolutionary 72-inch reflecting telescope, the nineteenth-century Irish aristocrat and avid astronomer William Parsons, third earl of Rosse, thought the whole chaotic mass looked just like a multilegged crustacean and thus christened it the "Crab." Some of the nebula's high-energy particles will take a long and winding journey through interstellar space and, in the far future, may finally impinge on Earth's atmosphere as a cosmic "ray."

Most stars are introverted souls, exhibiting few eccentricities during life or death. Only a small fraction can be considered to be celestial extroverts, yet these minority members of the stellar community captivate our attention the way a celebrity will overshadow the man in the street. About once every thirty or fifty years, a colossal star somewhere in the Milky Way explodes with a vengeance, just like the Crab nebula's progenitor in 1054. Earthlings don't get to see many of these spectacular explosions because, as with novae, most are considerably dimmed by a thick curtain of galactic dust and gas. Historical records note only five supernovae that have appeared in Earth's sky during this millennium, the last one in 1604. We seem to be overdue.* Bereft of detectable

*We are overdue no longer. Since this book was first published, a supernova exploded in the southern hemisphere. See the epilogue for further details.

supernovae in progress nearby, astronomers must either search for sudden brightenings in distant galaxies (they spot about ten extragalactic supernovae a year) or conduct autopsies on dozens of old supernova remnants in our galaxy—diffuse shells of gas largely detected with x-ray and radio telescopes—to reconstruct the event. Theorists, in the meantime, including the pioneering Hans Bethe of Cornell University, turn to their computers to unravel the inner workings of a stellar explosion.

Just a few handfuls of theorists, however, even attempt to model the dynamics of a star about to go supernova, since only a small number of sophisticated supercomputers in the world have been able to handle the complex calculations (though this situation will likely change as supercomputers become more commonplace). Astrophysicist Stan Woosley, whose office is set amid the stately redwoods that shroud the Santa Cruz campus of the University of California, is part of this close-knit group of researchers. His reason for pursuing this particular line of work is rather simple: "I happen to like explosions," he says spiritedly. Purists refer to them as "nuclear instabilities."

Some of Woosley's computer simulations, which trace the life of a star up until the very moment of its explosion, are programmed with 20,000 lines of code. And since the much-in-demand supercomputers have to work intermittently on a variety of problems, just one run through his maze of equations—seven hours of computer time in all—can often take six months to complete. Patience is a much practiced virtue among modelers of a massive star's demise.

Such computer simulations have shown that once a massive red-giant star begins to accumulate iron within its core, death is inevitable. The star faces its waterloo. "The temperatures become so extreme at this point, around ten billion degrees, that the iron literally starts to disintegrate," explains Woosley. "The iron breaks down into helium, and eventually into neutrons and protons." This tearing apart of the iron nuclei robs the star's core of precious energy, the very energy that kept the core intact. It's as if all the iron atoms had their chairs suddenly pulled out from under them. Cecilia Payne-Gaposchkin aptly described it as the time when stars "can no longer support themselves in the style to which they have been accustomed."

It is a critical moment. In less than a second, the star's hidden core, which encompasses a volume about the size of the Earth, collapses to the size of a thriving metropolis. In the blink of an eye, a neutron star is formed. "But you still have the overlying layers of the still-intact star raining down at very high velocities," points out Woosley. "Bethe and others have determined that when this material hits the dense compact cinder, it compresses the hard sphere to densities higher than an atomic nucleus. Therefore, like a coiled spring, the core bounces back." It is this rebound which generates the violent explosion. The recoil produces a tremendous shock wave that swiftly moves outward from deep inside the star. Over a day's time, this shock front, moving at many thousands of miles per second, works its way to the surface of the vastly bloated red giant and blows off the star's outer envelope. "A flood of elementary particles called neutrinos, that have been spewing from the core," adds Woosley, "may also help push the shock wave outward." In computer simulations, the shock wave gets bogged down for a second, but the neutrinos provide an added boost to keep the blast wave moving. Bethe jokingly refers to this crucial moment as "the pause that refreshes."

The shock wave is so powerful that, as it courses through the stellar material, it causes light elements to fuse into heavier ones. This is the second major way that elements heavier than iron get manufactured, rarer species such as gold, silver, and uranium.

There are actually two types of supernovae. White dwarfs are not entirely excluded from all the mayhem. If a white dwarf in a binary system should somehow accrete too much gas from its neighboring companion (enough to push it past the Chandrasekhar limit), the added weight will heat the dwarf's insides to a critical temperature, sparking a wave of fusion to run through the star. Once the fuse is lighted, it cannot be stopped. The dwarf becomes a thermonuclear powderkeg. Its once-quiescent carbon core, pressed into ignition, suddenly "burns" into oxygen; the oxygen immediately fuses into neon and magnesium; the magnesium into silicon; the silicon into iron—all in a matter of *one* second. "The resulting explosion blasts the white dwarf to smithereens, leaving nothing behind," says Woosley.

Some suspect that a bright supernova recorded by the Danish astronomer Tycho Brahe in 1572 was of this type. No pulsar has been discovered in the center of the shell-like remnant of the Tycho supernova. The last and greatest of the naked-eye observers, Tycho was also an arrogant man, who wore a metal nose after getting his own cut off in a duel at the age of nineteen. Tycho's careful analysis of the *Nova Stella,* as he called it, severely strained the widely held belief, initially expressed by the ancient Greek philosopher Aristotle, that the heavens were perfect and immutable.

Oddly enough, the brilliant light put out by a white-dwarf supernova originates, in large part, not from the detonation, but rather from the radioactive decay of the fusion by-products; the radioactivity causes the surrounding gases to fluoresce. This type of supernova, says Woosley, is the primary way in which the galaxy can receive an occasional supplement of iron. Neutron stars hoard most of their iron, but exploding white dwarfs can spread the metal generously. "The galaxy," points out Woosley, "is just a huge ecological system. Our iron-rich planet probably owes its very existence to the death of many binary white dwarfs, long extinct."

X-ray space telescopes have seen the expanding shell of matter from the Tycho supernova plow through the interstellar medium, heating up the gases to temperatures of millions of degrees. Eventually, this ever-growing bubble will strew the dwarf's elemental ashes far and wide. Physicist William Fowler, who received a Nobel Prize in physics for his five decades of accelerator experiments at Caltech's Kellogg Radiation Laboratory verifying the nuclear reactions that can take place inside a star, likes to say that "each one of us and all of us are truly and literally a little bit of stardust."

Other supernova remnants, such as the Crab nebula, will also make contributions. In time, the only reminder of the Crab's former glory will be the crushed and spinning cinder it leaves behind. Currently, the Crab pulsar is in the bloom of adolescence, pirouetting thirty times a second. Spinning at such a fast clip enables the powerful pulsar-generator to emit not only radio waves, but more energetic visible and x-ray waves as well. What looks like an ordinary star in the middle of the Crab nebula is actually the pulsar rapidly blinking on and off. Such vitality, however, will not last forever. Like a spinning top, the Crab pulsar is relentlessly slowing down. With each passing year, one rotation takes 15 millionths of a second longer to complete. By processes not yet fully understood, this lost rotational energy is somehow being transformed into radiant energy—just enough to keep the nebula brightly glowing in radio, visible, and x-ray radiation.

The Vela pulsar, which is embedded in an old supernova remnant whose wispy tendrils stretch across the southern sky, is three times slower than the Crab pulsar. The 20,000-year-old neutron star manages to emit a pulse every tenth of a second. Over the millennia, the supernova debris around Vela will disappear altogether. Vela will then be a naked pulsar, like Bell's LGM 1, that beeps even more infrequently. About ten million years after their birth, these cosmic beacons are probably heard from no more.

Since radio pulsars are essentially loners, the information astronomers can glean from the gyrations of these solitary neutron stars is limited. "With single stars, astronomers can't take advantage of Newton's laws. All we can do is silently watch them," explains Harvey Tananbaum, head of the high-energy astrophysics division at the Harvard-Smithsonian Center for Astrophysics. "But by observing a *pair* of stars orbit one another and determining their velocity and orbital period, we can begin to answer such questions as, 'How much do the stars weigh?' A binary system is a wonderful laboratory in which to make measurements. If a person wants to find out about an object, they pick it up, shake it, and feel it. Of course, we can't go out to a star and do this, but in a close double-star system, matter being transferred from one star to the other pokes and prods the system for us." It was not until the flowering of x-ray astronomy in the sixties and seventies, however, that neutron-star specialists obtained such a useful tool. Moreover, once acquired, it enabled astronomers to realize that the neutron-star stage, supposedly the deathly end for certain massive stars, can in reality be the beginning of great celestial upheavals, a twilight's last gleaming.

X-ray astronomy, which examines one of the most energetic regions of the electromagnetic spectrum, seemed doomed at the start. Soon after World War II, using surplus German V-2 rockets to loft instruments high above our x-ray-absorbing atmosphere, astronomers measured the X rays emanating from the Sun. Though very intense from a layman's point of view, the solar output of X rays is relatively meager by cosmic standards. Extrapolating from these solar measurements, theorists figured that x-ray emissions from faraway stars, not to mention galaxies, would be rather minuscule—not appreciable enough to warrant top-priority study. But one particular rocket flight in the early 1960s dramatically changed this situation.

It happened one minute before midnight on June 18, 1962. Riccardo Giacconi, Herbert Gursky, Frank Paolini, and Bruno Rossi had mounted a Geiger-counter-type detector onto a small Aerobee rocket and launched it from the White Sands Missile Range in southern New Mexico. A scant 350 seconds later, as planned, the rocket plunged back to Earth. It was one of the most fruitful six minutes in astronomical history. The field of x-ray astronomy had taken its first fledgling steps.

Giacconi and his colleagues had been trying to detect X rays from the Moon, generated as the energetic solar wind struck the lunar surface. The x-ray spectrum, it was surmised, might help them determine the Moon's composition. No lunar X rays were found during that flight, but the researchers hardly despaired. During the brief venture, the rocketborne detector had noticed a huge flux of x-ray radiation emanating from a region of the sky in the direction of the constellation Scorpius. Giacconi, a native of Italy and a guiding force behind the infant field, had long conjectured that extrasolar X rays would be racing about the heavens, radiating from such objects as supernova remnants and hot giant stars. The Moon experiment, funded by the U.S. Air Force, was an opportunity to start looking. But even Sco X-1, as the enigmatic source was later labeled, blazed with an intensity beyond anyone's imagination. Its x-ray emissions were tremendously more powerful than our Sun's.

A few years later, optical astronomers only added to the mystery: When they pinpointed the visible object supposedly disgorging this torrent of x-ray energy, it turned out to be a faint and variable, yet very innocent-looking, blue star. What could be the true nature of a star that emits a thousand times more energy in X rays than our own Sun radiates over all wavelengths? It was still a few years before Jocelyn Bell would notice her "bits of scruff," but Herbert Friedman of the U.S. Naval Research Laboratory, the rocket astronomer who first detected X rays from the Sun, suspected that a neutron star might be involved. Definitive proof, however, eluded him.

Throughout the sixties, the budding field of x-ray astronomy moved forward in fits and starts. Rocket and balloon experiments by a number of groups, each flight ever too brief, managed to uncover only a few dozen more sources. Tananbaum, a Ph.D. candidate at MIT during these pioneering days, clearly remembers chasing many an equipment-filled high-altitude balloon across the wide-open plains of Texas in pursuit of data for his thesis. This "horse-and-buggy" era ended when x-ray astronomers placed their detectors on earth-orbiting satellites that could leisurely scan the x-ray sky for months, even years, at a time.

The first x-ray-detecting satellite (SAS-1 or Small Astronomy Satellite-1) was appropriately nicknamed *Uhuru,* the Swahili word for "freedom." With the launching of *Uhuru* from a floating platform three miles off the coast of Kenya on December 12, 1970, the seventh anniversary of Kenyan independence, x-ray astronomy was being released from the confines of Earth's obscuring atmosphere. *Uhuru* and subsequent x-ray satellites succeeded in raising the number of known x-ray sources from about thirty to more than a few hundred. This new branch of astronomy blossomed fully with the 1978 launch of NASA's High Energy Astrophysical Observatory-2, which was quickly dubbed the *Einstein* observatory in honor of the centennial of the great physicist's birth in 1879.

Before *Einstein,* astronomical x-ray detectors basically acted like gigantic light meters, telling scientists which areas of the x-ray sky were "bright" and which regions "dark." The *Einstein* observatory, on the other hand, could actually focus the extremely short wavelengths and form a picture of the source, much like an optical telescope. Over its two-and-a-half-year lifetime, orbiting 300 miles above the Earth's surface, *Einstein* eventually relayed more than 7,000 images to its headquarters at the Harvard-Smithsonian Center for Astrophysics in Massachusetts. A team of x-ray astronomers is still deciphering the pictures and anticipating the construction of a more advanced space x-ray telescope that is planned to be one hundred times more sensitive.

With their newfound x-ray vision, astronomers perceived a disquieting quirk in the universe's temperament. Tycho Brahe dared to question the immutability of the heavens; x-ray astronomers shattered the tranquility. "The sixties were an unbelievable era for astronomy," says Tananbaum, who participated, under the leadership of Giacconi, in the development of both *Uhuru* and the *Einstein* observatory. "First, there was the discovery of the quasars; second was the discovery of radio pulsars; and third was the discovery of x-ray sources. All these findings led to a radical change in our thinking about the universe. Previously, we assumed that the universe was a place where stars changed only over millions, even billions, of years. We knew that there would be the occasional supernova or stellar flare, but generally celestial objects behaved like ladies and gentlemen. Now, we've come to understand that the universe is filled with explosions. Violence is the catch phrase."

It was soon after the launch of *Uhuru* that astronomers began to clearly see that the x-ray sky is a highly variable one. Along with the broad swaths of X rays surrounding supernova remnants and the energetic emissions from distant "exploding" galaxies, the satellite detected more than three hundred sources within the Milky Way that did not shine with a steady light, as visible stars basically do. Instead, these "x-ray stars" twinkled, flared, pulsed, and sputtered, like some sort of perpetual pyrotechnic display. Drastic variations in the x-ray luminosities of these objects occurred within days, hours, even minutes. Initially, it was a bewildering spectacle. "Most of us in the early days of x-ray astronomy were physicists by training," notes Tananbaum. "Instrumentation was our forte, not astronomy. On the one hand, it was an advantage: Riccardo Giacconi determinedly looked at the x-ray sky, even after everyone told him he wouldn't see a thing. But it was also a drawback: When we did find some peculiarity, we didn't have much of an astronomical lore to help us interpret the finding."

In 1971, a prominent x-ray source in the southern constellation Centaurus, known as Cen X-3, provided the much-awaited and crucial key. By examining the x-ray signal emanating from Cen X-3 in greater and greater detail, the *Uhuru* team came to see that its X rays were regularly turning on and off every 4.8 seconds. Cen X-3 was an *x-ray pulsar!* But this was a pulsar very different from the radio variety discovered by Bell. Every two days, for a dozen hours, the pulses would virtually disappear. This was a sure sign that Cen X-3 was not alone, but rather part of a double-star system. Evidence soon mounted that Cen X-3 was a tiny neutron star that periodically went in and out of view as it whipped around a very close and massive companion.

There was additional evidence to back this up: The pulsing varies smoothly in frequency. The timing of the pulses increases a bit whenever Cen X-3 comes toward Earth during its orbital journey and decreases as it moves away, just the way the whine of a fire engine sounds higher when the vehicle races toward you and then turns lower as it recedes. Almost overnight, the maddening twinkling of the x-ray sky was translated: Most bright x-ray sources in the galaxy, like Cen X-3 and Sco X-1, turned out to be neutron stars in binary systems.

"We knew something was in the air as the information on Cen X-3 first came in," says Tananbaum. By crudely plotting some of the data by hand at home one night, Tananbaum was the first to notice that Cen X-3's signal could abruptly change in intensity. The ragged-edged graph paper, now more than a decade old, still hangs on the back of his office door, like a parent's proud display of a child's nursery-school artwork. He honestly admits that he and his colleagues didn't immediately associate the cutoff with the eclipse of a star in a binary system, simply because they "didn't have a sufficient background in astronomy. We weren't thinking binaries at all!" For the *Uhuru* team, this was on-the-job training in basic astronomy.

The way in which Cen X-3 generates its X rays is a variation on a very familiar theme: stellar thievery. Soviet theorists broadly outlined the process before *Uhuru* was even launched. But, at that time, the idea was only one of dozens of x-ray-star models making the rounds in scientific journals. *Uhuru*'s data furnished the proof.

It is believed that Cen X-3 is drawing gas away from its companion (a gigantic but otherwise normal star) and funneling the material toward its two magnetic poles. In some ways, Cen X-3 acts like a radio pulsar in reverse: Matter is being drawn to the neutron star, instead of being expelled. "The gas, mostly hydrogen and helium, is falling onto a very compact object. That means it falls and falls, reaching speeds close to the speed of light," explains Harvard x-ray astronomer Grindlay. "As the matter comes crashing to a halt on the surface of the star, it gives off a tremendous amount of energy."

In fact, dumping blobs of matter onto a superdense object, which has a monstrous gravitational field, can release ten to twenty times more energy than if that matter were totally consumed in a thermonuclear explosion. Tossing a pebble from your hand onto the Earth kicks up a little dust; tossing that same pebble onto a neutron star releases more energy than the bomb that leveled Hiroshima. In the case of Cen X-3, this energy is released into space as a torrent of X rays. As Cen X-3 spins, its polar "hotspot" comes into view, and spaceborne detectors register an x-ray pulse.

X-ray astronomy has proven quite adept at puzzle-solving. The x-ray source Hercules X-1, for example, was found to be coupled to a flickering star called HZ Herculis, whose irregular light output had long been a mystery. "If I were an astronomer in 1910 and looked at HZ Herculis, I would have seen a moderately-hot F star," notes Grindlay. "But two nights later, if I looked again, I would have seen a hotter A star. Turn-of-the-century astronomers would have had difficulty understanding that this star is getting 'roasted' by the X rays coming off an invisible, compact companion. The stellar atmosphere is heated to such an extent that, as the neutron star orbits around, it makes the disk of HZ Herculis, the part facing us, periodically appear bluer, hotter, and brighter."

By the mid-1970s, astronomers discovered that neutron stars can be delightfully inventive in their antics. Regular radio or x-ray pulsing is not their only means of expression. For Grindlay, this revelation came in late 1975 as he was perusing a page from a computer printout, which listed some data gathered by two x-ray detectors mounted on the Astronomical Netherlands Satellite then orbiting the Earth. Two unexpectedly high numbers on the sheet piqued Grindlay's interest immediately, for they indicated that an x-ray source near the center of a globular cluster known as NGC 6624 had released, in a matter of one second, an intense burst of X rays that discharged more energy than our Sun emits over an entire month. "Imagine thousands of trillions of hydrogen bombs going off all at once," suggests Grindlay. "That might give you an idea of the strength of that burst. My first reaction was surprise and disbelief."

He immediately checked to see if the numbers were simply mistakes that had cropped up as the satellite data was shunted all over the world. Initially, the information had been radioed to a ground station in Chile, then sent over telephone lines to Maryland, and finally relayed to Madrid and Germany for final processing and transcription onto the magnetic tape that was mailed to the Center for Astrophysics in Massachusetts. But all the millions of digital bits checked out. The event was real. Within months, additional x-ray "bursters" were spotted by other research teams. Presently, around thirty are known; they are preferentially located in the Milky Way's central bulge, with a significant number occurring in the globular clusters that surround the galaxy. Having something in the cosmos pop off like a flashbulb was not a totally unfamiliar phenomenon. In 1973 the Vela satellites, put into

Earth orbit by the United States to spot any clandestine nuclear-bomb tests, detected bursts of gamma rays originating from random points in the sky. The galaxy, it seems, has a very tempestuous personality.

Sparked by Grindlay's discovery, astronomers excitedly examined x-ray bursters in greater detail, especially with x-ray detectors mounted aboard several satellites. The findings have led theorists to conclude that x-ray bursts are simply the neutron star's highly energetic rendition of a nova. X-ray bursters and white-dwarf novae, like Nova Cygni, are actually kissing cousins.

The reason that these particular neutron stars occasionally burst instead of pulse is probably owing to the fact that they have weaker magnetic fields. Under these circumstances, as the neutron star strips its binary companion and wraps a swirling disk of gas around itself, the matter is not dropped solely onto the neutron star's magnetic poles, as in the case of Cen X-3, but instead gets gently spread over the entire surface of the city-sized orb. The neutron star soon finds itself immersed in a layer of gas several yards deep. Squeezed to ever-hotter temperatures by the pressing force of gravity, the layer finally ignites "and bang, off it goes," says Grindlay.

"If the Sun did this, Earth would be frazzled, if not evaporated," adds Santa Cruz's Woosley, who has worked out many details of this burster model. "Even astronomers, who are used to dealing with powerful events, still get impressed by these objects."

"Ashes" from the explosion are blown off the stellar surface and form a sort of hazy, miles-deep atmosphere around the tiny neutron star. X-ray bursts are different from novae in one respect: Where white dwarfs need some 10,000 years to generate another explosion, x-ray bursters can generally refuel and pop off again in a matter of several hours. Indeed, one neutron star can send off volleys of X rays even more rapidly, like some kind of cosmic machine gun, as often as every thirty seconds. It's been dubbed a "rapid-fire" burster. In this situation, researchers believe, blobs of gas from the neutron star's accretion disk are intermittently "dripping" onto the star's ultradense surface, releasing enormous amounts of energy as the blobs crash upon the stellar surface. White dwarfs in binary systems do this too, although the results are much less spectacular. Astronomers are coming to suspect that, in certain binary systems, if a white dwarf should accrete just the right amount of material from its companion, it might even be able to collapse into a neutron star—without blowing either itself or its companion apart. Along with the explosion of massive stars, this could be another way that neutron stars are formed in the galaxy.

By peppering x-ray pulsars and bursters throughout the galaxy, nature has provided scientists with a ready-made laboratory in which to make accurate measurements of a neutron star's physique. As Tananbaum pointed out, a neutron star can be "weighed" by being part of a binary system. The precise pulses put out by the x-ray pulsar serve as the scale. The smoothly varying changes in the frequency of the pulse peg the orbital motion of the neutron star; the amount of time the tiny star is eclipsed behind its larger buddy provides a means of judging the size of the companion. With these pieces of information in hand, the mass of the neutron star can be calculated. The result: In the binary systems measured so far, neutron stars consistently weigh in at around 1.4 solar masses, the Chandrasekhar limit itself. "This is telling us something very fundamental about the formation of neutron stars," says Grindlay. "If borne out by further observations, it means that a certain amount of matter always collapses, while the rest is blown off."

Meanwhile, x-ray bursts have proved useful as tape measures. By watching how the x-ray waves released in a burst are "stretched" as they flee from the intense gravitational pull of a neutron star, astronomers can determine the width of the supercompressed object. It turns out to be, as predicted from theory, a span of about a dozen miles.

And, as with radio pulsars, glitches in the x-ray pulsing allow physicists to probe the inner sanctum of the star. In the prime of its life, an x-ray pulsar doesn't slow down; to the contrary, its rotation actually increases as it siphons material away from its companion, much the way a playground whirligig will speed up as you give it a push each time it goes around. Over one three-year period, the rate of Cen X-3's pulsing increased by six one-thousandths of a second. "This can give you an idea of what's inside," points out Tananbaum. "If it's a solid body, the star will respond one way. If it has fluids inside it, it will tend to 'slosh' and spin up in a different manner." Mass, size, and structure—all from the tick of a celestial clock.

Some might think it a paradox, but it is in old age that a star often puts on its most stunning performance, thus attracting the most attention. For the vast majority of stars, like our rather prosaic Sun, the grand spectacle will be the ejection of a glowing planetary nebula, leading to gradual decline as a white dwarf. If a star is lucky enough to be closely paired to another sun, an occasional nova might temporarily resuscitate the waning fire.

The most massive members of the Milky Way family, only a few percent of the stars in our galaxy, will mark their exit with a brilliant explosion—a gala end to a spectacular stellar odyssey. In some cases, these supernovae announce the debut of dense neutronic cinders that, for a while, can create a bit of havoc in the sky with their x-ray and radio flickerings.

But, just as Chandrasekhar opened the eyes of astronomers to possibilities of extinction beyond the white dwarf, might there be a stellar demise beyond the neutron star? Theorists started pondering this question half a century ago, and tantalizing evidence to back up their musings finally emerged in the reams of x-ray data radioed to Earth from *Uhuru*.

If a stellar core, its fuel exhausted, begins to shrink and its weight is more than two or three times the mass of our Sun, the collapse probably will not stop at the neutron-star stage. Instead, the stellar corpse will get sucked into a space-time abyss that has come to be known as the *black hole*.

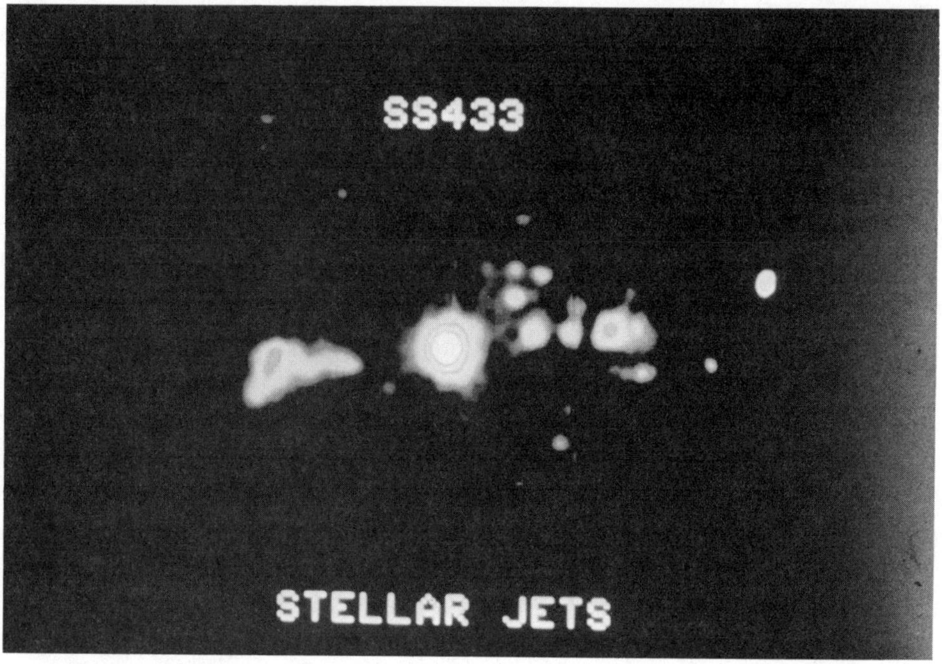

Do black holes truly exist? SS 433, a unique object in our galaxy, is a prime suspect. Allegedly, an unseen black hole lying in the center of this image, taken with an x-ray space telescope, is stealing gas from a stellar companion and expelling the excess in two oppositely directed jets. *Courtesy of Harvard-Smithsonian Center for Astrophysics, Cambridge, Massachusetts.*

3

THE ULTIMATE STELLAR CORPSE

> Ahead of them a blackness was eating the sky. . . . He was dimly aware that they must have crossed the event horizon. The line where things vanished forever—time and space together.
>
> —Alan Dean Foster, *The Black Hole*

The dawn of the 1970s brought with it a phenomenon that took many astronomers by surprise. It was during this decade that a never seen celestial object stepped off the pages of theoretical astrophysics journals and moved into the limelight, captivating the public's imagination. An audience that had never pondered the workings of a star and expressed little interest in red giants and nebulae took absolute delight in reading about the arcane physics of a creature known as the black hole.

"People love to hear about black holes because they're a tremendous threat, the 'Darth Vaders' of astronomy," says Douglas Eardley of the Institute for Theoretical Physics at the University of California at Santa Barbara, one of many theorists who have contemplated the mysteries of these alleged collapsed stars. The enthusiastic public response was sparked by astronomy's own doings: Throughout the seventies, data gathered by advanced x-ray, radio, and optical telescopes began to suggest, though indirectly, that a star's shrinking to utter invisibility could be more than a bizarre solution to a set of exotic mathematical equations. An idea once discussed by only a select group of specialists took on real substance. Following in the footsteps of the white dwarf and neutron star, the black hole has become a credible member of the universe . . . perhaps the strangest of all.

Though the current concept of a black hole is grounded in the most advanced theories of physics, the seed of the idea was planted more than two centuries ago. In 1783, as Great Britain was recovering from its war with colonial America, an English scientist and church rector named John Michell read a paper before the Royal Society of London suggesting that if a heavenly body got big enough, its gravity would become so powerful that not even light could escape from its surface. Arthur Eddington, writing on this hypothetical situation a century-and-a-half later, described the light rays as "falling back to the star like a stone on the earth."

To an observer nearby, such a star would be completely indiscernible, a black void in space. According to Michell's calculations, this celestial sleight-of-hand would happen if a star were 500 times wider than our Sun and just as dense throughout.* About a dozen years later, the great French mathematician Pierre Simon, Marquis de Laplace, arrived at a similar conclusion. "They had the wrong theory of light and the wrong theory of gravity," says black-hole expert Kip Thorne of the California Institute of Technology. "Yet, when they combined the two, they got the right prediction."

But Michell's "black hole" was hardly the genuine article. "Invisible star" would be a more apt description. The modern-day version—the awesome gravitational pit that figures in so many science-fiction stories —had to await 1915 and Einstein's General Theory of Relativity, which extended and reshaped the law of gravitation enunciated more than two centuries earlier by Isaac Newton.

*No star like this can exist. There are supergiant stars hundreds of times bigger than our Sun, but most of their material is spread very, very thinly.

Newton firmly established his scientific reputation for all eternity by noticing that the attraction a falling apple has for the Earth is the same attraction that keeps planets and comets in orbit around the Sun. He figured out that, in both cases, a universal force of gravity that can be precisely calculated is at work. But even Newton's brilliant thesis did not reveal the true nature of this pervasive force of attraction that keeps our feet on the ground. He quantified the effect but never explained its origins. In the Newtonian picture, all one can say is that matter somehow puts out feelers of force that keep planet to sun, sun to galaxy, and galaxy to cluster. That's not necessarily a failing. In everyday life, it's quite satisfactory. Newton's Universal Law of Gravitation is still adequate to insert most satellites into Earth orbit. Newton's law is not wrong; it's an excellent approximation in our mundane, everyday surroundings. The failure of Newton's model and the success of Einstein's are more readily apparent when gravity becomes very, very strong. Moreover, it was Einstein who audaciously revealed the "why" behind gravity.

Many who fall under the spell of Einstein's vision feel a little like Alice waking up in Wonderland. Suddenly, the universe looks "curiouser and curiouser." Matter turns out to be frozen energy ($E=mc^2$); objects shrink as they accelerate toward the velocity of light, the speed limit of the universe (special theory of relativity); and gravity is no longer a mysterious force, but rather a curvature in the four dimensions of space-time. Space, Einstein was telling us, is like a boundless rubber sheet, and large masses, such as our Sun, indent this flexible mat, causing any passing rocket, planet, or light beam to follow the natural depression. Celestial bystanders perceive it as an attraction to the Sun and call it gravity. According to this scheme, the planets are not held in orbit by invisible tendrils of attraction instantaneously flung out by the Sun; they simply travel along the warp that the Sun has created in this pliant sheet of space-time. The more massive the object, the greater the indentation.

General relativity was a geometric interpretation of gravity, and it was verifiable. "Almost at first sight, physicists around the world recognized general relativity as the most beautiful set of physical laws ever formulated," says Thorne. "So beautiful, in fact, that it was hard to believe that it could possibly be wrong."

Their intuition served them well. Einstein predicted that beams of starlight, which happen to graze the Sun, would get bent by a slight but detectable amount owing to the dimpling of space around the Sun—twice the bending postulated from previous suppositions. After World War I, British astronomers eagerly mounted two expeditions to prove Einstein's conjecture. On May 29, 1919, from sites in northern Brazil and the tiny island of Principe off the coast of West Africa, the researchers photographed some stars that were located near the edge of the Sun during a total solar eclipse, a rare event whose shadowy visage was visible from those tropical regions on that particular day.

To everyone's delight, the gravitational deflection of the stellar rays was well within the range of Einstein's calculation. The results certainly didn't agree with Newton's theory. More sophisticated versions of the 1919 experiment continue to this day. One method conceived by Irwin Shapiro, director of the Harvard-Smithsonian Center for Astrophysics, employs radio astronomy. He and others have confirmed general relativity's deflection predictions to within a few percent by using networks of radio telescopes to note how the strong radio emissions of pointlike quasars, the energetic nuclei of very distant galaxies, shift their position when passing near the Sun.

Eddington, a member of the 1919 expedition to Principe, called his participation in the verification of Einstein's Theory of General Relativity the greatest moment of his life. Banesh Hoffmann, who collaborated with Einstein in the 1930s, suggested in his biography of the illustrious scientist that "the eclipse observations gave the theory an aura of mystery and cosmic serenity that must have caught the fancy of a war-weary public eager to forget the guilt and horror of World War I." The name *Einstein,* to the physicist's great puzzlement, became a household word as soon as the results were announced.

Even as Eddington and the other Britishers were planning their enterprising journeys to the tropics, theorists began toying with Einstein's equations (as ivory towerists are prone to do) to see how the new law of gravitation operated in certain theoretical situations. In 1916, only a few months after Einstein introduced the basic principles of general relativity, the German astronomer Karl Schwarzschild embraced the idea and asked what would happen, gravitationally, if all the mass of an object (such as the Sun) were squeezed down to one point—a condition of zero volume and infinite density that mathematicians label a "singularity."

Shortly before contracting a fatal disease while posted at the Russian front during the war, Schwarzschild discovered that around this hypothetical point a spherical region of space could be defined out of which nothing—no signal, not a glimmer of light or bit of matter—could escape. This ethereal boundary has come to be known as the "event horizon," because no event that occurs within its borders can be observed from the outside.

The event horizon, it should be stressed, is not a solid barrier, like the ultrahard surface of a neutron star, but rather a gravitational point of no return: Once you stepped inside that invisible border, there would be no way out, only a sure plummet into the singular abyss at the center —a one-way ticket to oblivion. Trying to get out would be like trying to swim upstream against a raging current. Even if you attempted to send a distress signal to the outside world, the light beams or radio waves would only get trapped by the intense gravitational field within the border and bend completely back. If the Earth collapsed to a point, it would have an event horizon with a circumference of roughly two inches. Any object that dared to enter that marble-sized realm would never emerge from its clutches. The mass of ten suns squeezed into a dot would create a horizon 120 miles around. As you can see, as more mass is concentrated in the center, the gravitational "sphere of influence" extends farther and farther outward.

Even though you wouldn't be able to see this event horizon, you might know if you were getting close to one. Going back to that analogy of space-time as a rubber mat, Schwarzschild's imaginary point indents the fabric so much that space actually curves in on itself. At a certain distance, just outside the horizon, any light beams that follow this warp would eventually encircle the entire horizon, like a satellite going around the Earth. Theoretically, that means you could gaze into the distance at this point in space and actually see the back of your head! Picture the photons of light, carrying the image of the back of your head, orbiting completely around the spherical event horizon until they return to their starting point and enter your eyes. Unfortunately, under the circumstances, this bit of gravitational magic may be a bit difficult to witness. The strength of gravity increases so rapidly as you approach the event horizon of a star's-worth of mass that the pull on your feet is much, much greater than the pull on your head. So, as you're admiring the back of your hairdo, you're being stretched to shreds as if you were on a medieval torture rack.

Schwarzschild's calculations were merely an academic exercise, until 1939 when J. Robert Oppenheimer and two of his graduate students, George Volkoff and Hartland Snyder, showed that our cosmos could well be churning out those "singularities" with their forbidding event horizons: Begin with a star that has exhausted all its fuel. With the heat from its nuclear fires gone, the star's core becomes unable to support itself against the pull of its own gravity (as we saw in the previous chapter), and the stellar corpse begins to shrink.

If this core is weightier than a certain mass (now believed to be around two to three solar masses), Oppenheimer concluded, the stellar remnant would not turn into a white-dwarf star nor even settle down as a tiny ball of neutrons, because once the material is squeezed to densities beyond three billion tons per cubic inch, the neutrons can no longer serve as an adequate brake against collapse. Oppenheimer and his junior colleagues calculated from Einstein's gravitational theory that the star would continue to contract indefinitely. It would be crushed to that inconceivable singularity, a place where *all* the current laws of physics break down and from which no radiation can escape. The last light waves to flee before the "door" is irrevocably shut get so extended by the enormous pull of gravity (from visible, to infrared, to radio and beyond) that the rays become invisible, and the star vanishes from sight. Thus was born the modern-day concept of the black hole: It was the ultimate stellar death.

Oppenheimer and company didn't use the term "black hole" in their work; that provocative phrase wouldn't be coined until a few decades later. Instead, they called it "continued gravitational contraction." But it was the same animal. Space-time around the star becomes so warped that the star literally closes itself off from the rest of the universe. "Only its gravitational field persists," wrote Oppenheimer and Snyder in the journal *Physical Review.*

It's possible that this horrific collapse is accompanied by an explosion; some believe the supernova remnant Cassiopeia A, for example, may have a black hole lurking in its center. Cas A's widely scattered debris, filled with heavy elements, certainly suggests that a giant star exploded in that region of the sky some three hundred years ago, but the clocklike beeps of a radio pulsar have never been heard to echo from its depths. Perhaps a black hole "lives" there instead.

For that matter, there's no firm theoretical reason, as yet, to say that a black hole can't form more demurely. "If the shock wave generated by the collapse of a stellar core dissipates on its way out of the dying star, then you have a failed supernova explosion," points out Santa Cruz's Stan Woosley. "You could end up with a neutron core in the middle of this massive star. Eventually, the dense neutron center is going to 'eat' the rest of the star and turn into a black hole. The star could just wink out of existence."

If collapse to a black hole occurs, not only space but also time, the "fourth" dimension, would get turned on their heads. This is because time slows down within a strong gravitational field, a natural outcome of general relativity that has been proved many times. When master atomic-clock builder Robert F. C. Vessot of the Harvard-Smithsonian Center for Astrophysics rocketed one of his extremely accurate timepieces 6,000 miles above the Earth's surface, where gravity has loosened its grip, he found that the clock ran a tiny bit faster: It would have gained one whole second over Earthly clocks if it had been left in space for 73 years. Conversely, we on Earth, pressed down by gravity, would be one second younger than the spaceborne clock by comparison.

Black holes, the mightiest gravitational sinkholes in the universe, carry this effect to the extreme. If a group of observers were miraculously able to sit on the surface of a collapsing star, just before it shrinks within its event horizon to become a black hole, they would see only a fraction of a second tick away on their watches. But meanwhile many, many eons would pass in the rest of the universe; someone watching the collapse from afar, removed from the immense and eerie gravitational pull of the newly forming black hole, would think those voyagers were taking forever to cross the collapsing star's event horizon. Relativity, in this situation, lives up to its name. During a conversation with Einstein, writer Ashley Montagu regaled the physicist with a popular joke concerning this paradox. It involved two men from the Bronx:

"What is relativity?" asks the first man.

"Supposing," explains the second, "an old lady sits in your lap for a minute, a minute seems like an hour. But if a beautiful girl sits in your lap for an hour, an hour seems like a minute."

"And this is relativity?" responds the first man.

"Yes," answers his companion. "That's relativity."

"And from this Einstein earns a living?"

Einstein, according to Montagu, was heartily amused and declared it one of the best explanations he had ever heard.

But in 1939, all these intriguing paper-and-pencil revelations from general relativity were merely a game, a fascinating mathematical game. Neither Oppenheimer, Volkoff, nor Snyder ever suggested officially that totally crushed stars actually existed. And following the lead set by the influential Eddington ("There should be a law of nature to prevent a star from behaving in this absurd way"), other astronomers and physicists were optimistic that a better understanding of the death of a star would some day reveal how total stellar collapse is prevented. In the 1930s, it was still somewhat rebellious to think that a star could be scrunched beyond the white-dwarf stage, much less disappear from the face of the universe. "Like the unicorn and the gargoyle," Thorne has remarked, "the black hole seems much more at home in science fiction or in ancient myth than in the real universe." So, the astounding Oppenheimer-Volkoff-Snyder analysis was put on the shelf, marked as a "theoretical oddity."

This laissez-faire attitude didn't really change until the 1960s, when noted theorist John A. Wheeler, practically single-handedly, led an effort at Princeton University to bring these curious solutions to Einstein's equations back to the field of physics and connect them to the real world. Teaching a course on general relativity forced him to confront the scary proposition explored by Oppenheimer and his associates: What would be the fate of a star that is so massive that it simply cannot sustain itself against total gravitational collapse? Wheeler was sure that completely collapsed objects had a reality beyond the equations in Oppenheimer's dust-covered journal articles. "Wherever there is a crack in the armor that covers the unknown, thrust in a spear, pry it open, and look in!" Wheeler has commented.

Wheeler is credited with breaking down one of the last psychological barriers keeping physicists from seriously tackling the problem: In 1968, he christened this nameless state of collapse. It was he who inventively dubbed it the black hole. "Physicists, like patients in a physician's office," he once wrote, "only really believe they know what their problem is when it has been given a name."

The time was ripe. Even though astronomers had come to accept the existence of white dwarfs within the first decades of this century, it wasn't until the late 1960s that neutron stars gained respectability with Bell's detection of radio pulsars. As a result of that discovery, the presence of terrifically compact objects in the cosmos at last became more palatable. Black holes, in fact, appeared to be the next logical step. It didn't hurt that the term "black hole" was a catchy phrase.

According to Wheeler, who later joined the University of Texas at Austin, the issue clamored for attention. To what form were these stars collapsing? This, he has stated, "remains the great problem. All signs indicate that it comes to a point at which there is no 'after'—nothing more. It is not only matter that is crunched. The space surrounding the matter collapses too. Time, above all, comes to the end of the road."

That's what the current laws of physics tell us. Some wonder whether this view will alter once the Einsteinian concept of gravity and the laws of physics at the submicroscopic level (quantum mechanics) are successfully joined, creating the long-sought unifying theory of quantum gravity. Just as Newton's law of gravitation fails in regions of strong gravity, general relativity breaks down when it tries to face the infinite forces and densities of a singularity. Quantum gravity may not. Physicists may find that a new state of matter comes into play on the tiniest of levels, perhaps over distances as short as 10^{-33} (a billionth of a trillionth of a trillionth) centimeter. Just what that new state of matter is, however, remains frustratingly unknown.

At Princeton in the early sixties, Wheeler and his graduate students had more basic concerns. Their effort to prove the existence of a black hole consisted of two key tests: The strange beast had to be born under conditions that can actually be found in the universe, and it had to be stable enough to survive. The black hole passed the first test with ease. "The initial conditions for making a black hole aren't very stringent," says Thorne, who studied under Wheeler during this black-hole renaissance. "You simply need a star that's several times more massive than our Sun. We see them all over the place."

Stability was a tougher criterion to meet. Thorne got his students to tackle this problem once he traveled to Caltech in 1966 and proceeded to build up one of the world's leading centers of black-hole research, where graduate students and postdoctoral scholars are daily guided through the intricate byways of Einstein's equations of relativity. Located in the old Bridge Laboratory building on the Caltech campus, these black-hole investigators don't worry about reserving telescope time; their study of the cosmos is strictly cerebral. Only an occasional visit to a computer breaks their musings.

Many physicists pinned their hopes of killing off the black hole on the idea that it would fall apart under the slightest perturbation. By the early 1970s, however, this avenue of escape was effectively cut off. Some wondered, for instance, whether a black hole could form if the collapsing star were not quite spherical. Suppose it had "bumps" on it. But Richard Price calculated that the bumps simply smooth out during the star's collapse. Using a supercomputer, Larry Smarr of the University of Illinois even showed that when you throw one black hole at another—one of the most cataclysmic events imaginable—all you get is a bigger, very stable black hole. From a theoretical standpoint, black holes appear to be inevitable.

Although black holes are often described as "bizarre" or "exotic," to astrophysicists they're rather simple creatures, especially when compared to other celestial objects. A neutron star, despite its extreme compaction, has an intricate "geology" reminiscent of the Earth's interior with its core, mantle, and outer crust. The Sun is just as complicated. Solar astronomers have to concern themselves with the Sun's temperature, chemical composition, magnetic field, rotation, sunspots, and solar wind. A black hole, on the other hand, can be completely described by three elementary parameters: its mass, its electric charge, and the rate at which it is spinning. It has no other distinguishing features. Whether the matter that was swallowed up by the black hole was hydrogen, uranium, or comic books makes no difference. We'll never know, and there's no way to find out. Because of a hole's inherent simplicity, theorists often like to say that "a black hole has no hair."

But do black holes truly inhabit our Milky Way? Obviously, astronomers cannot photograph a black hole, but observers are capable of detecting the gravitational havoc a hole can wreak upon its surroundings. Many have compared a black hole to the Cheshire Cat in *Alice's Adventures in Wonderland.* The Cheshire Cat fades away until only his mischievous grin remains; the black hole leaves its gravitational effects to haunt us. Black holes can pull material apart, shred it to pieces, and gobble it up, releasing enormous amounts of energy in the process. It would be through these turbulent interactions, its effect on the matter in our galaxy, that a black hole would give itself away. The devastation would be especially noticeable if the black hole were a member of a binary-star system.

That's exactly what astronomers began thinking in the early 1970s when they took a closer look at the celestial shenanigans of a powerful x-ray source known as Cygnus X-1. Cyg X-1 was one of the first bright regions spotted in the x-ray sky during those pioneering rocket flights in the 1960s, which lofted sensitive x-ray detectors far above Earth's atmosphere. Later, the satellite *Uhuru* confirmed that this luminous x-ray-emitting patch in the constellation Cygnus was a particularly puzzling object. While other x-ray sources, Cen X-3 for example, turned out to be neutron stars that emitted regular pulses of energetic radiation, Cyg X-1 underwent sporadic variations in its x-ray intensity; there was no discernable pattern to its erratic flare-ups. Sometimes it flickered over periods as short as milliseconds. Whatever was giving off those X rays had to be a fairly small object.

Clever detective work by a number of radio and optical astronomers eventually determined that Cyg X-1's powerful X rays come from a double-star system, where a giant blue star (with the rather prosaic tag HDE 226868) is coupled with a dark, invisible companion. Regular variations in the optical spectrum of HDE 226868, a star of approximately 25 solar masses, indicate that the blue supergiant closely orbits its mysterious companion once every 5.6 days. The tempo of the orbital dance tells astronomers that the unseen companion must have a mass six to fifteen times that of our Sun. Could it be just an ordinary star? Some believe that a normal star of that mass would not only be visible, it would be ill-equipped to put out a lot of X rays. A compact neutron star doesn't fit the description either; neutron stars simply can't exist with weights beyond three solar masses. Applying a good dose of Sherlock Holmesian logic to this chain of circumstantial evidence, many astronomers have concluded that Cyg X-1 has to be a stellar black hole. It became astronomy's first prime suspect.

At least one Chinese astronomer has speculated that Cyg X-1 formed fairly recently. Historic records in China specifically note the sudden appearance of a new "star" in Cyg X-1's neighborhood, southeast of the bright star Vega, during the autumn of 1408. The stellar apparition was described as yellow in color and "as big as a lamp." This could have been the telltale glow of a supernova explosion, where black holes might be forged.

If we could hover above Cyg X-1 today, astronomers imagine that we would witness a whirlpool of enormous dimensions. Observation as well as theory suggests that Cyg X-1 is pulling matter off of its generously endowed companion and forming a disk of gas around itself that is being flattened by both centrifugal and gravitational forces. The material does not fall straight into the black hole, but instead orbits the space-time drain in ever-tighter spirals. Wheeler once compared it to traffic converging upon a sports stadium from all directions and becoming more and more tightly packed as the cars approach their destination.

As the gas gets squeezed tighter and tighter, its temperature rises accordingly. Heated to tens of millions of degrees, the hot gases proceed to emit copious amounts of x-ray energy. According to the scenario, this is the radiation that x-ray telescopes observe before the matter is sucked into the abyss and lost to our view. It may take weeks or even months for any one blob of gas to travel the few million miles from the outer edges of the disk to the point of no return. In its last moments, the gas swirls around the hole thousands of times each second, possibly causing those rapid x-ray fluctuations.

Claims of detecting a black hole are so fraught with peril (and possible ridicule) that it was many years before astronomers announced that they had found another candidate, this time in the southern sky. Throughout the early 1980s, Anne Cowley of Arizona State University and two colleagues, David Crampton and John Hutchings of British Columbia's Dominion Astrophysical Observatory, often traveled to the Cerro Tololo Inter-American Observatory in Chile to identify the visible counterparts to many x-ray sources dotting the Large Magellanic Cloud, a satellite galaxy of the Milky Way located some 180,000 light-years from Earth and named after the sixteenth-century explorer

Ferdinand Magellan, whose crew first sighted the cloud on their historic voyage around the globe. As expected, most of the sources in the cloud seemed to be associated with binary-star systems whereby a tiny, unseen neutron star siphons material from a more normal and visible companion, generating a flood of X rays along the way. But one highly variable x-ray source, LMC X-3, didn't quite match those specifications. Interpretation of its properties was a little more troublesome. The visible star in the LMC X-3 system, a blue giant of about 5 solar masses, was whipping around its x-ray-emitting companion particularly fast—a little too fast. "We thought there had to be something wrong with our observations," recalls Cowley.

But additional measurements in the fall of 1982 with Cerro Tololo's 4-meter telescope only confirmed the unusual motion. By Cowley's figuring, the giant blue star in the LMC X-3 system is frantically circling an invisible companion once every 1.7 days; the two are separated by only 11 million miles. Applying the laws of orbital motion, the dark, compact companion turns out to have a mass of about ten of our Suns—strong ammunition for those who think it's the black-hole remains of a giant star that was completely crushed by gravitational collapse. "There's no other way to hide a star that's more massive than the visible star we see," points out Cowley. "Normally, when we go out to an observatory, we sit around the dinner table and talk to the other astronomers about what we're doing. But, that time, we didn't dare say anything until we got home and double-checked our calculations." Many consider this to be the most convincing black-hole candidate to date, since the distance to the system and its orbital motions can be determined more accurately than in the case of Cygnus X-1. Agreement among astronomers on either, however, is far from unanimous.

Some astronomers enthusiastically argue that a black hole also powers a unique celestial "sprinkler" called SS 433. In this binary system 15,000 light-years distant, a tiny unseen object is continually stealing gas from a huge companion, wrapping an accretion disk around itself, and then ejecting some of the matter in two oppositely directed jets. In an x-ray picture taken by the *Einstein* observatory, the jets are seen to extend outward for 500 light-years (3,000 trillion miles). Opinion is split over the nature of the engine creating these awesome jets. Some contend it's a neutron star; others, like Harvard-Smithsonian's Jonathan Grindlay, favor a black hole, since some evidence indirectly suggests that the unseen object weighs about ten solar masses. "It's likely that matter is being drawn toward the black hole so fast," he conjectures, "that the excess gas gets squirted outward along the north and south poles. The black hole just can't swallow it all." And like the lazy wobble of a top spinning on a smooth surface, the black-hole accretion disk (or neutron star) appears to be precessing around its polar axis, causing the two jets to slowly move back and forth in a 164-day cycle.

SS 433 is like no other object in the galaxy. It may represent a snapshot in time—a brief evolutionary phase that other x-ray binaries experience at one time or another, perhaps lasting for ten thousand or a hundred thousand years. Cygnus X-1 may have gone through this jet-stage already, or may in the future. SS 433 is an especially attractive object to extragalactic astronomers, because it provides a miniature laboratory, "close up," for studying a process that may be energizing the hearts of some distant galaxies and quasars, where narrow beams of material are jettisoned over distances of a *billion billion* miles (see chapter 6). SS 433 expert Bruce Margon of the University of Washington has said that "it could serve as a prototype for gaining an understanding of the violent extragalactic events that are among the greatest mysteries in astronomy."

"Such phenomena as Cygnus X-1 and SS 433 can be explained by black holes, but the evidence is still circumstantial," cautions Thorne. "We can't be more than eighty percent sure, since we don't have a definitive signature, like the pulses from neutron stars—something jumping up and saying 'I am truly a black hole.' That will likely come from detecting gravity waves"—a means of observing the universe's most enigmatic, and violent, goings-on.

Even as this century's most celebrated physicist was developing his famous theory of general relativity, he realized there would be peculiar side effects to its picture of gravity as a curvature in space-time. Einstein's equations revealed that if a mass were suddenly accelerated or jostled to and fro, it would produce ripples in that pliable "sheet" of space-time, similar to the way electrons moving up and down an antenna generate radio waves in the air. But while electromagnetic waves travel *through* space, gravity waves are an actual disturbance in the fabric of space, comparable to the waves that propagate outward from a stone dropped into a pond. Detecting these ripples is an accomplishment that has eluded celestial observers so far, but no one seriously doubts their existence. There's a small chance that this space-time rippling can occur when you bang your fist on a table or jump rope. Even the Earth emits very feeble doses of gravity waves during its perpetual journey around the Sun. But only the most awesome cosmic events create any appreciable waves. Particles, planets, and stars caught in the path of such a wave would experience space itself contracting and expanding.

Such a "spacequake" would provide astronomers with an entirely new form of information about the universe. Visible and infrared light, radio waves, and X rays—radiation gathered by current instrumentation—are emitted almost entirely by individual atoms, molecules, and high-energy particles. Gravitational waves, by contrast, are generated by the bulk motions of huge amounts of matter, objects that are vibrating, collapsing, or exploding. More important, these periodic distortions in the structure of space-time can blithely pass through interstellar dust, planets, and galaxies as if they weren't there. Nothing can absorb them to any great extent. Thus, gravity-wave astronomy would offer a totally new window on the universe. Its penetrating power may allow astrophysicists to observe cosmic processes that, at least for now, can only be imaged on a computer graphics terminal—from the last millisecond gasp in the life of a star to the titanic collision of two black holes.

It's an intriguing, even frightening prospect but, in Einstein's day, hardly of much consequence. He was sure they would never be seen, and with good reason: Imagine that an aged star has just died somewhere in the Milky Way, compressing its core into either a neutron star or black hole and shedding its outer envelope in a brilliant supernova explosion. The collapse is so sudden that the star's gravitational field undergoes a rapid transformation, and it is this change which is transmitted toward Earth at the speed of light as a gravitational wave. But by the time this warp in space-time hits the piece of paper you are now reading, alternately stretching and shrinking the paper, the energy will have been spread out over a vast distance. This will cause it to be so weak that it will change the sheet's dimensions by, at best, only one-thousandth the diameter of a proton—less than a ten-thousandth of a trillionth of an inch. No wonder Einstein considered it a hopeless cause.

Waves rushing in from the Virgo cluster of galaxies, some 50 million light-years away, would be a thousand times weaker. Such minuscule movements make gravity-wave astronomy look virtually impossible, but amazingly enough, physicists around the globe are now gearing up to record these tiny quivers in space-time.

The first attempt at snaring the ghostly ripples was made in the 1960s when Joseph Weber of the University of Maryland, the acknowledged father of gravity-wave astronomy, built the first set of detectors. It was Weber who devised a nifty technological trick for trapping a wave: A string of gravitational waves passing by, he surmised, would ever so slightly squeeze a huge solid cylinder in and out like an accordion. But then, like a tuning fork, the bar would continue to "ring" long after the waves departed. This ringing would be the gravity waves' calling card—vibrations that could be monitored by electronic sensors attached to the bar.

With massive aluminum cylinders operating on both the Maryland campus and the Argonne National Laboratory near Chicago, Weber announced in 1969 that he had registered some pulses, bursts that appeared to originate in the center of our galaxy. Catalyzed by the announcement, several groups quickly constructed their own detectors. Excitement within the physics community was, however, short-lived. Though a few facilities have reported seeing several pulses similar to Weber's, most have failed to detect the same kind of ringing. "What Weber was seeing remains interestingly unknown," says William Fairbank, who oversees one of the world's leading gravity-wave detection projects at Stanford University in California. But Weber's relentless effort did transform a seemingly hopeless endeavor into a lush, new field of experimental physics.

Even as Weber was operating his first detector, Fairbank and colleague William Hamilton were designing a supercooled version to reduce random thermal noises in the bar and thus increase their chances of picking up the fainter, gravity-induced movements. Two of these advanced gravity-wave "telescopes" currently reside in the very heart of Stanford's campus, within a cavernous room that was once part of the original Stanford Linear Accelerator Laboratory. Another one has been built at Louisiana State University under the direction of Hamilton. Each detector consists of a hulking metal tank, a sophisticated thermos bottle, actually, that encloses a solid several-ton aluminum bar cooled with streams of liquid helium to within two degrees of absolute zero (−460 degrees Fahrenheit). The rumblings of passing cars and trucks are damped by suspending each phone-booth-sized metal bar in a vacuum from special springs. Not even a gentle hammering on the outside shield generates a response. And, similar to Weber's original scheme, electronic devices positioned on the end of the bulky aluminum cylinder convert its minute movements into electrical signals that are recorded and scrutinized for a gravity wave's unique fingerprint. As of this writing, one of the frigid Stanford bars can sense a shiver as tiny as 0.0000000000000001 inch (a few thousandths the size of a proton), a world's record for the field.

The Stanford group is not alone; this is truly a global pursuit. Other supercooled bars of varying designs and materials (including niobium, sapphire, and silicon) either have been or are being assembled at the Universities of Maryland, where Weber continues his pioneering work, Rochester, Rome, Tokyo, and Western Australia, as well as in China and the Soviet Union. Like surveying instruments, an array of detectors distributed over four continents will enable this worldwide community of gravity-wave astronomers to more precisely pinpoint the direction from which a wave is coming and to rule out local glitches. A signal received more or less simultaneously at several detectors thousands of miles apart would be strong evidence that a gravity wave has passed through our solar system.

But watching bars of metal or crystal vibrate is not the only means of stalking these alleged ripples in space-time. By the 1970s, researchers such as Robert L. Forward of Hughes Research Laboratories and Rainer Weiss of the Massachusetts Institute of Technology recognized another way to catch a gravity wave: Attach mirrors to three heavy masses, suspend the masses in a vacuum some distance from one another, and monitor their relative motions with a laser beam to see if a passing gravitational wave has wiggled the weights.

A gravity wave acts by compressing space in one direction while expanding it in the other, so a popular configuration for such a laser detector is an L shape, with a mass at each end and one at the corner. "Imagine a gravity wave coming straight down on the L," suggests physicist Ronald Drever, head of Caltech's gravitational physics group.

"Then the masses in one arm will draw closer together by a distance many times smaller than an atomic nucleus, while the other two get farther apart. A millisecond later, as the wave passes by, the effect will reverse." Test models of this setup have been built in Scotland, West Germany, and at MIT, and are under development in France and Russia. But Caltech presently boasts the world's largest gravity-wave laser antenna. Each arm of their L stretches 40 meters out, a little more than 131 feet. Upon entering their annex on the Caltech campus, you might think you had stumbled upon the campus utility room. Most of the space is taken up by what looks like two long water pipes meeting at right angles. They're actually the evacuated pathways for the laser beams that are used to measure the distances between the system's 22-pound brass test masses.

What do gravity-wave astronomers hope to see with such intricate instrumentation? At present, an array of Stanford-type bar detectors and Caltech-style laser systems could conceivably detect a nearby supernova popping off in our Milky Way. And, according to Thorne, the gravitational pulse could be almost as revealing as a motion picture of the collapse itself. "There's just no other way for us to see this," explains the Caltech theorist, "since the electromagnetic waves emanating from the core are completely absorbed by the outer layers of the star. Gravity waves, on the other hand, can travel from the stellar heart with impunity." The wave pattern might even reveal the stellar core of the supernova bouncing for a brief moment—squishing down to a pancake and then stretching into a football before settling down.

But there's a catch. Astronomers estimate that such stellar explosions, and the resulting collapse of the remnant cinders into ultradense neutron stars or black holes, occur in our galaxy only once every 30 or 50 years. Even improving gravity-wave detectors to "see" out to the Virgo cluster of galaxies would provide only a few events a year—small reward for such a complex enterprise. "But the universe always turns out to be more complicated than we originally think," Fairbank quickly counters. "Remember, people almost didn't want to look for X rays in space, believing they just weren't there."

The same may be true for gravity waves. If a city-sized neutron star develops a blemish on its surface (an inch-high "mountain," perhaps), it will continually transmit gravitational waves as that bump rapidly spins round with the star's rotation. Encouraged by an intriguing discovery in radio astronomy, theorists have come to suspect that a whole circus of gravity-wave emitters lurk in the heavens. In 1974, Russell Hulse of the University of Massachusetts at Amherst and Joseph Taylor, now with Princeton University, revealed that our galaxy harbors a most interesting stellar couple: a radio pulsar, which flashes seventeen times a second, swiftly orbiting another neutron star. It was the first radio pulsar found to have a companion. Having such a precise clock circling another star, about once every eight hours, was a relativist's dream. This rare type of binary, free of interfering gas, provided the first demonstration, albeit indirectly, that celestial objects can indeed emit gravitational radiation. By using radio telescopes to closely time the pulsar's clock over several years, Taylor and colleagues were able to show that the two neutron stars are inexorably moving closer to one another. With each passing year, their orbital period steadily decreases by about one ten-thousandth of a second, bringing them roughly one yard nearer to one another. That's just the change expected if the binary system is losing energy in the form of gravity waves. "There's no other known mechanism to bring these two stars together," says Taylor.

Current gravity-wave detectors are too crude to sense this continuous, weak emission of gravity waves. But when those miles-wide balls of compact matter finally spiral into one another, likely merging to form a black hole, they'll release a sizable burst. The clash of the Hulse-Taylor pair won't occur for another 300 million years, but gravity-wave detectors proposed for construction in the 1990s and twenty-first century, laser systems with miles-long arms or instruments placed out in space, for instance, might detect this type of event in other galaxies, out to a billion light-years. "At that distance," notes Thorne, "it's reasonable to expect an event once a week." And if neutron stars can orbit one another, it's not too difficult to imagine that two black holes can form a binary pair and collide as well. "That's my favorite event in terms of the physics that could be learned," says Thorne. "If I had to lay bets on how we'll finally get one hundred percent proof that black holes exist, it would be that event."

Two black holes orbiting one another would eventually spiral in, releasing a unique set of gravitational waves: first, an ever higher-pitched whine during the final minute of the fateful twirl, then a cymbal-like crash as the holes coalesce, and finally a ringing-down as the merged bodies settle down. A future generation of gravity-wave antennas, presently on the drawing board, have a chance at seeing these monstrous collisions as far as the edge of the visible universe. With such a wide vista, astronomers might see several a day. An air of skepticism always surrounded the idea of neutron stars until new techniques, the collection of celestial radio and x-ray waves, arrived on the scene to confirm their existence. Likewise, black holes may need to make some gravity-wave telescopes quiver before they are fully accepted.

Just as electromagnetic radiation comes in all sizes, from radio waves to gamma rays, gravity waves will certainly vary in length, depending on the source. Collapsing stars, for example, are expected to send out ripples that stretch some hundreds or thousands of miles from peak to peak. But if a supermassive black hole residing in the very center of an active galaxy gobbles up a star, it might send out waves hundreds of millions of miles in length. These very long wavelengths will best be detected by spaceborne systems with extremely long arms.

Actually, some tentative gravity-wave tests have already been made far beyond Earth's atmosphere. For six days in March 1980, scientists from the Jet Propulsion Laboratory and the radio astronomy department at Caltech monitored the radio signals flowing to and from the *Voyager I* space probe during its sojourn to the outer planets. If a gravity wave had rolled by, it might have varied by about a millimeter that vast distance between our planet and *Voyager I,* and this would have shown up as an advance or delay in the phase of the spacecraft's radio signal beaming back toward Earth. Some hope to also measure in this way the general background of gravity waves out in space. Such a murmurous hubbub is more important than it sounds. Our explosive beginning, the Big Bang that occurred some 15 billion years ago, may have emitted a burst of gravitational radiation that echoes through the universe to this day. The *Voyager I* test didn't find it, nor have tests with Pioneer spacecraft, which have been tracked for three weeks each year since 1981. But more sensitive attempts are scheduled to be made as new interplanetary probes are launched into space.

As gravity-wave astronomers were assembling their heavy cylinders and laser chambers, the black hole itself took on a new look. British theoretical physicist Stephen Hawking startled the astrophysics community in the 1970s when he suggested that black holes may not be so "black" after all: that they can emit energy in the form of elementary atomic particles.

Hawking, who has made formidable contributions to astrophysics despite a severe disease of the nervous system that keeps his paralyzed body confined to a wheelchair and makes his labored speech impossible to comprehend without an interpreter, discovered this unsettling effect while pondering how a black hole might affect its surroundings from the viewpoint of an atom. Space-time gets so twisted near a black hole that it enables pairs of particles (a nuclear particle and its antimatter mate) to pop into existence just outside the event horizon. You might think of the extreme conditions as allowing energy, extracted from the hole's intense gravitational field, to be spontaneously transformed into matter. At times, one of the particles will disappear into the black hole, while the remaining one flies away. As a result, the hole's *total* mass-energy is reduced a smidgen. In a way, that means the black hole is actually decaying or evaporating. "Like everyone else at that time," Hawking has said of his finding, "I accepted the dictum that a black hole could not emit anything. I therefore put quite a lot of effort into trying to get rid of this embarrassing effect. It refused to go away, so that in the end I had to accept it."

For a regular black hole, one formed from normal stellar collapse, this bizarre quantum-mechanical process is just about meaningless. A black hole with the mass of a few suns would need more than 10^{67} (10 million trillion trillion trillion trillion trillion; a *one* followed by 67 *zeros*) years to shrink away to nothingness in this manner. From a human's perspective of time, that's practically everlasting. But Hawking pointed out that the universe may have manufactured another type of black hole in its distant past: infinitesimally small black holes, some with event horizons only 10 one-trillionths of an inch around. Multitudes of these miniature black holes could have been forged during the crushing and turbulent first moments of the Big Bang. If so, the very tiniest will have vanished by now; but objects containing the mass of a mountain, that were compressed to the size of a proton a fraction of a second after the universe's birth, would be shedding the last of their mass at this very moment.

Like a ball rolling down a hill, the evaporation of such a "mini" black hole accelerates as time progresses. The more mass the tiny primordial black hole loses, the faster and faster the evaporation proceeds, until it finally reaches a cataclysmic end: The hole dies in a final and violent burst of gamma rays, powerful enough to supply humankind's total energy needs for several decades. So far, astronomers have not detected any signals in space that exactly resemble these gamma-ray explosions. This might mean that conditions during the Big Bang were not conducive to forming primordial black holes, or created very few. Just in case, astronomers are keeping their sensors tuned for that distinctive *pop*.

Because the modern-day version of the black hole—its physical characteristics and calculated gravitational effects—emerged from the General Theory of Relativity, it's interesting to ask whether Einstein's equations hinted at other unusual and unheard of "creatures." The answer to that question is yes, but with many strings attached. "While solving the equations of relativity for the most extreme cases—that is, when gravity becomes very, very strong—mathematical physicists found a veritable zoo of bizarre objects," admits Thorne. "By the time I was a graduate student at Princeton University in the early 1960s, not only black holes, but also white holes, wormholes, and tunnels-through-hyperspace were all on the scene—honest-to-God exact solutions to Einstein's equations. We were trying to keep this quiet, because we had no reason to believe that any but the black hole existed in the real universe."

White holes are simply black holes in reverse; instead of an event horizon, white holes have an *anti*event horizon, a region of space into which nothing can enter. Like cosmic geysers, white holes are only able to spew matter into the universe. For a fleeting moment, several physicists seriously considered the idea that quasars might be white holes.

Wormholes are even stranger entities. In this mathematical solution, the indentation of Schwarzschild's singularity in the flexible mat of space-time becomes so deep that it punches through "hyperspace," opening up passageways to not only remote corners of this universe but other universes as well—perhaps even into our own distant past. Tunnels-through-hyperspace are a variation on this theme. "Tunnels, like black holes, have event horizons," explains Thorne. "But, whereas a black hole has an all-encompassing singularity, a tunnel has a ringlike singularity." Some have speculated that a well-aimed plunge through the ring's center would take you to those other universes and other times.

Science-fiction fans were ecstatic. Science seemed to be handing them an outlandish yet legitimate means of jumping around both space and time. Such well-worn plot devices as hyperspace drives and time machines began to appear more plausible. It was a means of avoiding the ultimate stellar crunch. "This speaks for something very deep in people," says general relativist Eardley. "It's an escape from death. You don't get killed after you fall into the black hole; you just come out somewhere else. It makes a great mythology."

And a myth it remains for serious relativists. By the 1980s, white holes, wormholes, and tunnels-through-hyperspace all failed miserably in thought-experiments aimed at proving their reality, tests that black holes passed with ease. While these very exotic celestial objects can exist on paper, they cannot survive in a real universe filled with light and matter. They fall apart under the slightest perturbation. If a hyperspace tunnel somehow did appear, for instance, it would be destroyed in a microsecond by stray radiation falling into it. One physicist has called such instability "a sign of the good taste built into the general theory of relativity."

"There's a solution to certain equations of physics which tells us that, at some moment in time, all the molecules in a room can have the proper velocity to go rushing through a vent, leaving us gasping for air," explains Thorne. "That's a real solution in physics, but no one really expects it to happen. Well, that's more likely to occur than a tunnel-through-hyperspace." The universe has retained its sanity. We no longer have to worry that our great-great-great-grandchildren could ever come back through some time tunnel and kill us, thus cutting to shreds the principle of causality. Fiction, for now, remains stranger than truth.

Thorne points out, though, that physicists have barely scratched the surface in understanding how gravity behaves when it is extremely strong. New exhibits in the cosmic zoo may yet come out of general relativity. But presently there seem to be only three possible routes to stellar extinction: Depending on its mass and environment, a star can either shrink to a white dwarf, collapse to a neutron star, or, rarest of all, cut itself off from the rest of the universe as a black hole.

The real importance of black holes to cosmic affairs, however, may lie in their ability to generate huge amounts of energy within a proportionately small region of space. Run by gravity power, black holes are potentially the universe's most powerful engines, especially when nature constructs them on a grand scale. Astronomers have come to speculate that certain kinds of black holes—gigantic specimens containing the mass of millions, even hundreds of millions of suns—could be occupying the centers of many galaxies, releasing enormous amounts of energy when properly "fed." Maybe, as we shall see in the next chapter, a supermassive black hole is even controlling some strange goings-on in the nucleus of our own Milky Way.

A frog's-eye view of the "mushroom patch." The Very Large Array, set on a New Mexican desert plain, is a *Y*-shaped collection of 27 dishlike antennas that serves as radio astronomy's premier eye on the Milky Way and the universe. *Photo courtesy of Rogier A. Windhorst.*

4

WRAPPED IN AN ENIGMA

> On a clear, moonless night in midwinter or midsummer, a plume of starlight rises motionless behind the scattering of constellations. . . . The Milky Way is our island universe—our galaxy. . . .
>
> —Charles A. Whitney, *The Discovery of Our Galaxy*

Even at 55 miles per hour, the desolate terrain seems to pass by in slow motion. Only an occasional group of piñon pine on the side of a hill or, farther off, the stark profile of an erosion-sculpted mountain breaks the monotony.

But suddenly, after driving over a rise on Route 60, a few dozen miles west of Socorro, New Mexico, the weary traveler will come upon a sight unlike any other on Earth: 27 dishlike antennas lined up for miles over the flat, desert Plains of San Agustín. Airline pilots who fly over the ancient, mile-high lake bed have dubbed this gigantic Y-shaped configuration the mushroom patch. But to astronomers it is simply known as the VLA—the Very Large Array that since the late 1970s has served as radio astronomy's premier eye on the universe.

Twenty-four hours a day, seven days a week, the VLA's majestic white dishes move in unison, like a mechanical version of the Rockettes, to gather the radio waves sent out by the universe's myriad celestial inhabitants. On one day, the antennas may trace the wispy outlines of a gaseous nebula to see how its molecules tumble and collide, leading astronomers to the birthplace of new stars. The next day, a VLA computer appropriately named the Boss may order the 82-foot-wide dishes to point toward a supernova and snap a "radio picture" of the debris racing away from the mighty stellar explosion.

More recently, astronomers have been training the multiple antennas on the very center of our galaxy, the Milky Way, to reveal both arc-shaped streams of radio-emitting plasma and a mysterious compact object that some suspect is a multimillion-solar-mass black hole. Thick clouds of interstellar dust and gas keep the sight well hidden from optical telescopes. Although theories regarding the true nature of these phenomena proliferate, experts are at least agreed that something extraordinary is occurring in the heart of our galaxy.

These new revelations concerning the structure of our galaxy took some Milky Way specialists by surprise, for they had believed for a while that the major components of the galaxy were fairly well established. Now, the Milky Way, once thought of as fairly quiescent, is unveiling a galactic nucleus filled with energy and motion. This discovery is the latest episode in one of astronomy's oldest endeavors: investigation of this pale, broad strip of light that snakes across the heavens.

In preindustrial times, when neither the murky haze of air pollution nor the glaring lights of urban sprawl obscured the nighttime sky, the Milky Way was a reassuring fixture in the heavens: a creamy ribbon of light, intertwined with sinewy patches of the deepest black, that encircled the entire celestial sphere. Sadly today for many city dwellers the existence of this "plume of starlight," as astronomer Charles Whitney described it, is simply a matter of faith. But to the eyes of ancient Egyptians and Greeks, the diffuse band, in which the solar system is immersed, looked very much like a river of milk. Thus, the term *galaxy* is derived from *gala,* the Greek word for milk.

To our great-grandparents, the Milky Way *was* the universe, a finite system of stars that was surrounded, perhaps, by a limitless void. In the nineteenth century, it was even daring to think that fuzzy, spiral-like spots of luminosity situated above and below the plane of the Milky Way might be other, more distant "island universes." Change has come swiftly, sparked by a series of improvements in astronomical instrumentation. Nowadays we speak with ease of a cosmos, filled with billions of galaxies, that has been expanding for ten to twenty billion years. This picture is so firmly entrenched in modern-day cosmologies that it's disorienting to be reminded that astronomers' perception of the universe's structure was vastly different as little as a century ago.

In many ancient societies, it was steadfastly accepted that the stars and Milky Way were features fastened to the surface of a crystalline sphere which was fixed some distance from the Earth and which completed one revolution every twenty-four hours. According to this scheme, the Sun, Moon, and planets were attached to smaller transparent spheres.

Ptolemy (Claudius Ptolemaeus of Alexandria), an astronomer and geographer of Greek descent, immortalized this view in the second century A.D. in his classic work entitled *Almagest* (Arabic for "the greatest compilation"). Strongly influenced by the philosophy of Aristotle, who lived five centuries earlier, Ptolemy proclaimed a universe that was flawless and unchanging, with a motionless Earth poised prominently in its center. The sky was an intricate maze of wheels within wheels that soundlessly revolved around God's central creation—Man. The Ptolemaic system is often ridiculed today, but it was basically a commonsense approach that successfully modeled the limited data of that pretelescopic era. Even today, when you gaze up at the sky at night, it can be difficult to dismiss the immediate sensation that you are caught in the middle of a gigantic hollowed-out vault whose surface is liberally sprinkled with flickering lights.

Ptolemy's Aristotelian vision of perfection held sway for some 1,400 years. It was accepted as the authoritative model of the universe throughout the Middle Ages. The Ptolemaic system began to crumble, however, by 1543. In that year, near death, Polish churchman and scholar Nicolaus Copernicus published his *De Revolutionibus Orbium Celestium* (On the Revolutions of the Celestial Spheres), a radical tome that asserted that the Earth and planets were in orbit about the Sun. "All this is suggested by the systematic procession of events and the harmony of the whole universe," wrote Copernicus, "if only we face the facts, as they say, 'with both eyes open.' "

Further cracks in Ptolemy's long-held theory appeared in 1609, once Galileo Galilei aimed a newfangled telescope toward the heavens and discovered a cosmos that was far from perfect. By putting a concave lens behind a convex one, the famous Italian experimentalist was readily able to see that the Sun's surface was peppered with blotchy dark spots, that the Moon was filled with jagged mountains and craters, and that the foggy glow of the Milky Way was actually comprised of a multitude of individual stars. "Upon whatever part of it the telescope is directed," wrote Galileo, "a vast crowd of stars is immediately presented to view. Many of them are rather large and quite bright, while the number of smaller ones is quite beyond calculation." An ancient Greek philosopher named Democritus* would have been pleased. Around the fifth century B.C., he had intuitively reasoned that the Milky Way was a dense array of tiny stars.

The universe was turning out to be not only blemished, but changing and evolving. By 1718 the English astronomer Edmund Halley (of comet fame) determined that several prominent stars—Arcturus, Betelgeuse, Sirius, and Aldebaran—had markedly changed their positions in the sky since the days of Ptolemy. Even earlier, keen-eyed observers had noticed that a star called Mira in the constellation Cetus, the Whale, varied its light output, periodically dimming and brightening. To the consternation of many, the universe was revealing a certain capriciousness: Its stars were scattered about at varying distances and were far from immutable.

With this new information came new attempts at fathoming humanity's true position in the cosmos. In 1750, an Englishman named Thomas Wright proposed that the universe might be arranged as a set of close concentric rings, much like the rings of Saturn. Anticipating twentieth-century discoveries, he suggested that the Sun was situated, not in the center of this circular disk, but off to the side. Wright, strongly guided by religious views, reserved the center as a source of spirit and life. More importantly, Wright was the first to explain how the Milky Way might

*This prescient man is best known for his belief, a couple of millennia before it was proven, that all matter consists of infinitesimally small particles called atoms.

be an optical effect: It appeared as a band, he said, because we observe the disk of the universe edge-on, the Earth and Sun being immersed in the flattened slab of stars. Wright would later reject this model for one he considered more theologically appealing, but not before a newspaper account of the theory, inaccurate as it was, influenced his contemporary, the eminent German philosopher Immanuel Kant, to write a book on the subject.

After several years of cogitation, Kant concluded that all the stars in the disk must be moving in large orbits around a central point; otherwise, according to Newton's law of gravitation, they would all gravitate toward one another, eventually disrupting the disk. "Perhaps it is up to future generations at least to find the area in which this central point of the fixed star system, to which our Sun belongs, lies," wrote Kant, "or maybe even to determine the precise location of this central body of the Universe, to which all parts are uniformly attracted." Kant guessed it might be Sirius, the Dog Star. He must have felt that the universe's "central body" had to be a special kind of object; why not the brightest star in the heavens? We now know that Sirius, a fairly ordinary A-type star situated only 8.7 light-years away, appears particularly bright only because of its relative nearness to Earth. There are millions just like it throughout the galaxy. But if Kant were with us today, he might not be too surprised to learn that those "future generations," in other words, today's astronomers, are coming to suspect that the Milky Way's center may indeed house an exceptional member of the celestial family.

William Herschel, eighteenth-century England's self-taught Prince of Astronomy, attempted to determine the Milky Way's structure in a more rigorous, scientific fashion than either Kant or Wright, an approach that introduced statistical analysis into astronomical study. Besides discovering the planet Uranus, hundreds of double-star systems, and thousands of nebulae, Herschel, with the aid of his devoted sister Caroline, spent many hours at the eyepiece of his homemade, 20-foot-long telescope with its 18-inch mirror meticulously counting the stars that were visible in over six hundred distinct regions of the sky. By assuming that the faintness of a star was an indication of its distance, he deduced that the galaxy looked very much like a gigantic convex lens, filled with millions of stars. Unaware of the Milky Way's obscuring dust and gas, which prevent us from seeing its bright center, Herschel concluded that the Sun was located near the hub of this disk, a belief that was widely accepted by the astronomical community well into the twentieth century. Even a more modern and mammoth survey, conducted by the Dutch astronomer Jacobus Cornelius Kapteyn at the turn of the century, seemed to confirm this positioning. In a way, it was a return to the Ptolemaic notion of cosmic superiority: The Sun and its attendant planets were privileged members of the galaxy, with front-row seats on all its activity. Toward the end of his life, Herschel had serious doubts that his telescopes had the power to fathom either the limits of the Milky Way or its true shape. In his later writings, he even considered the possibility, as some Greek and Roman philosophers did, that the Milky Way was an unlimited system of stars that extended into an infinite space. But initially, Herschel figured that the diameter of the luminous disk stretched about 850 times the distance to the bright star Sirius (about 7,400 light-years from end to end). That's less than a tenth of the current estimate of the Milky Way's extent, but a considerable dimension for its day.

It would take a sharp-eyed farmboy, born and bred in the heart of America's Midwest, to banish the solar system from its regal position as center of the known universe and place it in a more rural location. As a teenager, Missourian Harlow Shapley worked as a cub reporter for the Chanute, Kansas, *Daily Sun* and entered the University of Missouri intending to major in journalism. A delay in the opening of the university's journalism school forced him to change his plans. In his reminiscences of that time, Shapley put it this way: "There I was, all dressed up for a university education and nowhere to go. 'I'll show them' must have been my feeling. I opened the catalogue of courses and got a further humiliation. The very first course offered was a-r-c-h-e-o-l-o-g-y, and I couldn't pronounce it! . . . I turned over a page and saw a-s-t-r-o-n-o-m-y; I could pronounce that—and here I am!"

Shapley went on to Princeton University in New Jersey for his graduate education, receiving a doctorate in 1913. Soon afterward, he was hired to work at the Mount Wilson Observatory, situated in the tawny and often barren San Gabriel mountains on the outskirts of Pasadena, California. This was a rather prestigious post for an up-and-coming astronomer, since the mile-high observatory then housed the largest light-collector in the world—a 60-inch reflecting telescope funded by steel magnate Andrew Carnegie—and was scheduled to install an even bigger 100-inch reflector in 1917 (now, sadly, shut down after nearly seventy years of use owing to the light pollution of the Los Angeles area).

For several years, Shapley used his time on the 60-inch telescope to study Cepheid variable stars in globular clusters. Globular clusters are round, tightly packed aggregations of stars that look, through a telescope, remarkably like fuzzy snowballs. More than a hundred of these dense celestial balls, many containing hundreds of thousands of stars, are scattered above and below the plane of the galaxy. Cepheid variables (so named because the first one discovered happened to be in the constellation Cepheus the King) are quite amazing stars; they dim and brighten with uncanny regularity. The period of these pulsations can range from half a dozen hours to over a hundred days. Shapley himself figured the phenomenon was caused by the Cepheid's outer envelope periodically surging outward and inward like some stellar tide. It's a relatively brief phase that certain stars undergo during their lifetime. A Cepheid is at the pinnacle of its brilliance when it swells, and dims as it shrinks. Polaris, the familiar North Star situated at the tip of the Little Dipper's handle, is a Cepheid variable whose brightness varies by a factor of about 10 percent every four days.

Several years before Shapley arrived at Mount Wilson, Henrietta Leavitt at the Harvard College Observatory discovered a wonderful link between a Cepheid's period and its luminosity: the longer the time it takes to go through one complete cycle of variability, the brighter the star. The Cepheid variables' period and luminosity were uniquely related. This characteristic made them superb beacons. In uncovering this relationship, Leavitt provided astronomers with one of their handiest yardsticks for measuring distances throughout the galaxy. Once they were calibrated, Cepheids became astronomy's most useful standard candles. Astronomers were soon able to pick out a far-off Cepheid star, note its period, and from this infer its true *absolute* brightness (the luminosity you would observe if you were essentially right near the star). The distance to the Cepheid immediately followed: By measuring the Cepheid's *apparent* brightness in the sky, a much fainter magnitude, astronomers could figure out how far away it must be to appear that dim.

By using this and other methods, Shapley was able to determine the distances to dozens of globular clusters. The results were somewhat disturbing to believers in the dominant Kapteyn model of the Milky Way. Shapley had found that the clusters were roughly distributed in a large spherical halo centered around a faraway point in the constellation Sagittarius. The young researcher boldly reasoned that this point must be the true hub of the Milky Way disk, not the region around the Sun. The new model put the Sun decidedly off to one side, about two-thirds of the way to the outer edge. His shifting the Sun from its pivotal location was a feat that could be matched only by Copernicus's earlier removal of the Earth from the center of the solar system. In 1917, at the age of 32, Shapley had finally placed us in our rightful position in the suburbs of the Milky Way. "The solar system is off-center and consequently man is too. . . . He is incidental," Shapley would remark many years after his discovery.

Astronomers were not immediately thrilled by Shapley's conclusion; the idea was strongly resisted in some quarters. But, in time, additional lines of evidence tended to support Shapley. By 1927, for example, the prolific Dutch investigator Jan Oort was able to demonstrate that stars in the galaxy were indeed rotating about a point lying in the direction of Sagittarius, the very same center around which Shapley's globular clusters were distributed. Our own Sun, located about 30,000 light-years from the galactic center, orbits the hub at a velocity of some 150 miles per second, completing one full circuit every 250 million years. The dinosaurs were just starting to plod through primitive Paleozoic forests when last our planet passed this way. *Homo sapiens,* it seems, is caught up in an ever-widening cosmic whirl: As the Earth spins on its axis, it also moves at 66,000 miles per hour about a star, which travels around a wheel-shaped galaxy, which glides onward through the outer edges of an expansive cluster of galaxies.

Estimates of the Milky Way's dimensions have been adjusted several times since Shapley's day. The young Mount Wilson astronomer had originally calculated that the entire galaxy was about 300,000 light-years in diameter, a size that greatly disturbed many observers. But his figure was considerably reduced once astronomers recognized that a thin haze of dust and gas permeates the Milky Way disk. The dimming of starlight as it travels through interstellar matter must be taken into account when making distance measurements. Because of intervening dust, Cepheids appear fainter and consequently more distant than they really are. Nevertheless, even the modified figures were a tremendous jump in galactic dimensions, at least when compared to Herschel's or Kapteyn's earlier estimates.

The 100 billion or more luminous stars in our galaxy are now believed to occupy a region of space that measures about 100,000 light-years across. That means a ray of light, traveling at some 186,000 miles per second, needs 100,000 years to journey the 600,000,000,000,000,000 miles from one end to the other. A rocketship traveling at space-shuttle speeds (approximately 17,500 miles per hour) would require four *billion* years to complete the voyage. Even at that, it's sometimes difficult to grasp the enormity of such a distance. If the entire solar system—out to frigid Pluto's orbit—were reduced 100 trillion times until it was roughly the size of this book and Earth was no bigger than a germ, the Milky Way would still stretch 6,000 miles across, the distance from London to Los Angeles. The star nearest to our Sun, which becomes a microscopic pinprick of light in this shrunken cosmos, would lie a quarter of a mile away.

Imagine being whisked far out into intergalactic space and observing the Milky Way from its side. From that vantage point, you might be amused to discover that our galaxy closely resembles the archetypal "flying saucer" found in countless science-fiction movies. At the center you would see a huge, round, yet somewhat flattened bulge, containing, predominantly, a dense concentration of old and reddish stars, probably formed fairly early in the history of the Milky Way; surrounding this bulbous nucleus would be the much thinner disk, a couple of thousand light-years thick, whose gas and dust are enriched with the heavy elements brewed in stellar interiors. The solar system is located here. The disk is also where the Milky Way gives birth to its newest members, in particular hot and massive blue stars. The disk and bulge, in turn, are completely immersed in a diffuse spherical "halo," which houses the galaxy's oldest stars. The halo is a sparsely populated region where many of the globular clusters circle the galaxy in eccentric orbits. One astronomer described the clusters as grouping themselves about the plane of the galaxy "rather like bees around a flower."

Globular clusters are particularly interesting conglomerates because they are celestial Methuselahs, the most ancient objects in the Milky Way galaxy. They contain no young stars and virtually no gas or dust out of which new stars can be hatched. Astronomers believe that these clusters were the first entities to be created as a vast primordial cloud, our nascent galaxy, began to condense. Thus, the rotund aggregations provide astronomers with vital clues to the manner in which the Milky Way came into existence. One can think of them as skeletal markers left behind as a roughly spherical blob of protogalactic gas, composed almost entirely of hydrogen and helium, collapsed over a period of a billion or more years to create the flat, luminous disk we inhabit today. "Globular clusters tell us what the galaxy was like more than ten billion years ago," explains Kitt Peak National Observatory astronomer Catherine Pilachowski, who specializes in the study of these clusters. "In some ways, they preserve the chemical characteristics of the gas from the original cloud out of which the Milky Way formed."

Because of this very fact, globular clusters have long been the focus of intense astronomical research. It has been determined that some globular clusters are as much as fifteen to eighteen billion years old, yet some (still controversial) estimates of the age of the cosmos suggest the universe is only *ten* billion years old. This creates a rather awkward situation. It's similar to the conflict that geologists and astronomers faced in the nineteenth century when evidence suggested that the Earth was older than the Sun. Such a dilemma is usually a strong signal that some basic astronomical assumptions will soon be revised. Whether revisions will be made to the models of stellar evolution used to date globular clusters or to the evidence that establishes the universe's age has yet to be settled. It is one of astronomy's most pressing concerns. "The globular clusters and the stars within them are just pinpoints of light to us on Earth," comments Pilachowski. "Yet, from those shreds of evidence, we will eventually piece together a magnificent construct."

With our solar system embedded inside the dusty plane of the Milky Way, discerning the exact configuration of the disk becomes a difficult task. It's almost like trying to determine the pattern on a piece of china with your eyes leveled on the edge of the plate. But, since the disks of many other galaxies in the universe display beautiful spiral structures, astronomers have long assumed that the Milky Way, too, has massive arms, laced with cosmic dust and gas, that wrap themselves around the galactic hub like coiled streamers.

By the 1930s, identifying the Milky Way's spiral arms became one of the most important items on astronomers' agenda. Without this information, observers probably felt as if the map of their hometown were missing. Oddly enough, World War II aided the effort greatly. Because of the fear that the Japanese might attack the United States' West Coast, the Los Angeles area was blacked out nightly during the conflict; the sky was darker than it had been for decades. While other staffers at the Mount Wilson Observatory were involved in war work, German-born Walter Baade was designated an "enemy alien" and restricted to the Pasadena area. With almost unlimited time on the 100-inch reflector, Baade got the chance to intimately study the Andromeda galaxy, at a distance of two million light-years the spiral galaxy closest to us. He pushed the telescope to its limits. In doing this, Baade came to recognize that, while old red stars tend to huddle in a spiral galaxy's bulge, young blue-white supergiant stars and bright gaseous nebulae are inclined to line up along a galaxy's spiral arms, like lights along an airport runway. He thought of them as "candles on a birthday cake." Thus, it was learned that certain objects make excellent spiral-arm indicators.

Applying this newfound knowledge to our own Milky Way galaxy, William Morgan and two student assistants at the Yerkes Observatory in Williams Bay, Wisconsin, painstakingly determined the distances to dozens of blue giant stars and luminous nebulae within the solar neighborhood. Morgan's view was far from complete, since it is difficult for optical telescopes to see beyond 20,000 light-years in the dust- and gas-filled plane of the Milky Way. "It's like taking a piece of an apple," comments University of Maryland astronomer Leo Blitz, "and trying to deduce the structure of the entire apple." Yet Morgan's analysis was a tremendous breakthrough. Segments of a few spiral arms could be traced from his data. When Morgan presented his results at a 1951 meeting of the American Astronomical Society, he received a rare, emotional ovation that included clapping of hands and stomping of feet. The conferees were enthusiastically responding to the long-awaited news that their "hometown" had at last been mapped—at least partially.

Within a year of Morgan's announcement, a spiraling pattern was confirmed and extended by the new kid on the block, radio astronomy. Radio telescopes provided researchers with a powerful means of penetrating the dust and haze that restricted Morgan's ground-breaking study. By tuning to hydrogen's distinctive 21-centimeter-wavelength emissions, radio astronomers in the Netherlands and Australia, two countries that quickly embraced the new technology, were able to trace multiple streaks nearly all the way around the galaxy, since concentrations of hydrogen are enhanced in spiral arms. Yet, even to this day, there are conflicting opinions as to how many spiral arms the Milky Way actually has and how far they extend. Astronomers are not certain because it is very difficult to unambiguously determine the distance to a particular clump of gas and so trace an arm exactly.

Are there two arms? Four arms? Educated guesses fill the astronomical journals. After mapping the distribution of atomic hydrogen gas in the outer regions of the galaxy (beyond the Sun's galactic orbit), Blitz and Berkeley astronomers Shrinivas Kulkarni and Carl Heiles surmised from symmetry arguments that there are four gently curving arms arranged around the galaxy's center and that they are splayed somewhat open. In support of this contention, giant molecular clouds appear to be fairly well concentrated along the lengths of these arms. One arm is labeled the Sagittarius-Carina arm and lies 6,000 light-years from the Sun, in toward the center of the galaxy. Notably bright clouds found along this arm are the Lagoon, Trifid, and Eagle nebulae. Another arm can be found 6,000 light-years away in the opposite direction, "behind us" toward the Milky Way's outer edge. This is the Perseus arm and is one of the appendages that Morgan first detected. In the configuration proposed by Blitz and his colleagues, a third arm called Cygnus grandly sweeps around one side of the galaxy, while a fourth unnamed limb remains essentially hidden behind the galactic center. Our solar system, it turns out, is not situated on any of these major segments but, instead, appears to be caught on the inner edge of a small loose branch known as the Orion spur; besides a certain rocky planet, third out from a star called Sol, the Orion nebula is the spur's most famous inhabitant.

Milky Way specialists agree that galactic geography is hardly an established field. At this stage, it can deal only in approximations. The maps that astronomers are now generating are akin to the first crude charts of the New World, which could only guess at the magnificent expanse of the North American continent and where much of the landscape was labeled *Terra Incognita.* The four-armed pattern still cannot explain all the complex structure observed in the inner Milky Way, which "may be grand," Harvey Liszt of the National Radio Astronomy Observatory once stated at a Milky Way conference, "but it might not be so pretty."

But why should there be spiral arms at all? The billions of stars and myriad interstellar clouds that make up the disk of a spiral galaxy like the Milky Way do not all move rigidly in step, the way a phonograph record revolves; the inner regions of the disk orbit the center much more frequently than the outer regions. So, why don't the spiral arms simply wind themselves up in an eon or so, like a fishing line on a reel, as the galaxy rotates around and around? For years, this "winding dilemma" had many astronomers scratching their heads in puzzlement.

The answer was self-evident once astronomers recognized that spiral arms are merely particular sections of the disk that are temporarily highlighted. The material in the arms is continuously changing. When looking at a picture of a gracefully spiraling galaxy, one has to remember that dim stars and dust-filled clouds, too faint to be photographed, are smoothly distributed in the vast, dark regions between the arms.

In the early 1960s, building on an idea introduced by Swedish astronomer Bertil Lindblad, mathematical scientist C.-C. Lin of the Massachusetts Institute of Technology and astrophysicist Frank Shu, then with the Harvard College Observatory, suggested that spiral arms simply mark the position of a density wave, a rotating region of compression that slowly moves through the flat disk of the galaxy. Unlike the disk of stars, this density wave does rotate rigidly, somewhat like the spokes of a wagon wheel. It's like a spiral-shaped sound wave rippling through the disk. This might explain why the largest molecular clouds (and likewise the youngest and most brilliant stars) tend to concentrate in the galactic arms. As the disk's gas, which travels faster, periodically passes through this compression wave during its rotation around the galactic center, the material gets squeezed, huge clouds form, and within several million years big new stars turn on to illuminate the spiral structure. "In some ways," remarks Lin, "it's as if the stars and gas are encountering a cosmic traffic jam." The highly luminous supergiant stars, which largely delineate the spiral lanes, are so short-lived that they die off by the time they move out of the traffic tie-up. Therefore, says Shu, the easy visibility of the giant stars allows astronomers "to mark the crest of a spiral density wave as clearly as whitecaps mark the crest of a breaking water wave." As the bright beacons die out and move on, they are continually replaced by fresh batches of stars that are newly cooked up in the shock zone.

How does a density wave get started? Lin compares it to the ringing of a bell: "When you hit a bell it develops a distinctive tone. Well, the same is true for disklike galaxies." Irregularities in the disk, he says, can cause it to "ring," only in this case the reverberation comes in the form of a standing spiral wave. How long this ringing persists is still a matter of debate. There is statistical evidence to suggest that spiral patterns can be long lasting, the exact lifetime varying from galaxy to galaxy. MIT's Alar Toomre, on the other hand, believes that a spiral pattern is fairly transient, arising primarily during such disruptive events as two galaxies passing by one another. "Isn't it curious," says Toomre, "that many of the nicest spirals have suspicious companions?" Toomre doubts that when our great-great-great...grandchildren a few hundred million years from now take a picture of the famed Whirlpool galaxy (one often used in advertising as the prototypical spiral) they will see the elegant curving arms that are so prominent today.

For many, the density-wave theory has successfully explained why certain spiral galaxies exhibit a beautiful "grand design": two or more roughly symmetric arms extending over great distances. Having such lengthy, coherent limbs means that a galaxywide process must be involved. But density waves are probably not the only mechanism at work. In fact, astronomers are coming to understand that various processes—sometimes acting separately, sometimes simultaneously—can contribute to a galaxy's spiraling nature. "Galaxies with well-constructed spirals arms are pretty," points out physicist Philip Seiden, "but many spiral galaxies are actually quite messy," appearing to be constructed of a multitude of disjointed curvy fragments.

In computer simulations conducted at IBM's Thomas J. Watson Research Center in New York state, Seiden and Humberto Gerola have shown that these more "feathery" spirals might arise as bursts of star formation propagate over distances of thousands of light-years. A chain reaction takes place: Supernova shocks and stellar winds in one molecular cloud induce star formation in a neighboring cloud, and so on down the line. Seiden compares it to a fire racing through a forest. "In some ways, I'm calculating the probability of one burning tree causing the neighboring tree to burn," says the IBM researcher.

The millions-of-years-long process can be capsuled on Seiden's TV-sized computer-graphics screen. On the green-toned monitor, a computer-generated galaxy slowly rotates at different rates along the disk, wrapping the chains of digitized stars around the galactic hub and thus turning them into spiral arms. "It's like putting milk into your cup of coffee in the morning," explains Seiden. "You stir it up, and spirals form." Their formation is inevitable, because the liquid in the center rotates faster than the liquid near the rim. Any one spiral arm in Seiden's model eventually fades away as its fuel is exhausted, but new ones are constantly created and bent back by the differential rotation. Whether the flocculent arms are tightly or loosely wound depends on the speed of rotation.

Space x-ray telescopes have discovered that the interstellar medium, the gas and dust between the stars, certainly does get churned up by supernova explosions as they successively pop off over millions of years like a lengthy string of sparklers. Recently, a "superbubble" of hot gas, heated to a temperature of a couple of million degrees, was spotted in the direction of the constellation Cygnus, right by a giant gas cloud known as the Great Rift. The bubble's roughly spherical volume stretches nearly 2,000 light-years in diameter and is expanding at a rate of 20,000 miles per hour. To give you an idea of the enormity of that volume, a few tens of thousands of Orion nebulae would fit into it comfortably. Theorists figure the Cygnus superbubble got inflated as several generations of massive stars within it exploded in succession over the last three million years, each blast wave plowing the shell of material at the boundaries of the bubble farther and farther outward and also eating into the surrounding rift to trigger a new round of star formation. It's likely that such bubbles are a common occurrence in the Milky Way. As much as ten percent of the galaxy's volume could be laced with these tunnels of hot rarefied gas. Whether such chains of supernova explosions play a major role in the creation of the Milky Way's spiral arms, however, has not yet been determined.

The world-renowned Milky Way specialist Bart Bok, who liked to describe himself as a night watchman of the galaxy, once remarked that he and his fellow Milky Way researchers were notably self-assured in the early 1970s because the size, mass, content, and structure of the Milky Way galaxy appeared to be reasonably well determined by the start of that decade. New challenges in the field seemed to be dwindling. Young graduate students, anxious to tackle only the hottest topics for their doctoral dissertations, thought of Milky Way research as a pursuit that was long past its prime. "Even many of the workers most active in galactic structure research turned their attention elsewhere," adds Blitz. "There was more to be learned, it seemed, from observations of other galaxies than from observations of our own."

Little did they realize that an explosion of data would swiftly rain down upon them, forcing Milky Way observers to refashion the galactic tableau. Astronomers soon learned, for instance, of the existence of those giant molecular clouds that line up along the galaxy's spiral arms like vertebrae in a spinal column—massive entities never before envisioned. At least a few long-kept secrets concerning star formation were readily unraveled shortly after the discovery of these gigantic complexes. The Milky Way became a "hot" topic once more.

By the early 1980s, some of the most tantalizing new data concerning the Milky Way was being gleaned from deep within the center of the galactic nucleus. Something—no one knows exactly what as yet—lies hidden in the bowels of our galaxy, behind the hazy glow of distant suns, and it is radiating surprising amounts of energy across the electromagnetic spectrum. In radio waves alone, the nucleus of the Milky Way broadcasts with an intensity several times greater than galaxies of similar size and type. This is a relatively new discovery. For decades, the core of our galaxy was modern astronomy's most impenetrable realm. Layer upon layer of dust and gas float in space between us and the Milky Way's center; this gaseous barrier reduces the visible light emanating from that region drastically, a trillion times over. That means for every trillion visible photons, or particles of light, being emitted from the galactic center, only one manages to avoid the clutches of an absorbing bit of dusty flotsam on its way toward Earth. Consequently, optical telescopes could be declared legally blind as they attempt to scan this murky, yet mysteriously active, section of the Milky Way galaxy.

"If you look at the world with monochromatic vision, it looks pretty dull," points out Caltech radio astronomer Kwok-Yung Lo, who has spent many hours of observing time probing the galaxy's inner domain. "But by gathering different wavelengths, we can see much better. As a result, the galaxy is becoming more and more interesting." As pointed out in earlier chapters, radio, infrared, and high-energy waves, such as X rays and gamma rays, have a penetrating edge over visible-light waves in that they can more easily pass through the diffuse interstellar medium (though not necessarily through Earth's much thicker atmosphere). Therefore, by using instruments that gather radiation emitted in these other regions of the electromagnetic spectrum, astronomers can bore directly through the dust and gas, right into the very heart of the galaxy.

Karl Jansky, with his gangling radio receiver, was the first to cut through the dusty obstacle course. The crackling "star noise" he gathered from the Milky Way's center (one 1930s reporter wrote that it sounded like steam escaping from a radiator) strongly hinted that the galactic nucleus was more than a home for aged red stars, though it took quite a while for the field of astronomy to follow up on this lead. A few decades elapsed before more advanced radio telescopes would map our galaxy's central landscape in much better detail. From this venture, astronomers are discovering that the galactic center displays a bewildering topography, one that has obviously experienced a complex evolution.

As radio astronomers proceed on their journey toward the galactic core, they first come upon the outer borders of the galaxy's central bulge, some 15,000 light-years from the center, and find it surrounded by a bevy of giant molecular clouds. Farther in, about 10,000 light-years from the core, they detect a turbulent ring of hydrogen that is not only rotating, but also moving outward at velocities of more than fifty miles per second. This could be a sign that our galaxy is a barred spiral; the motions can be interpreted as gas flowing under the gravitational influence of a bar-shaped concentration of stars lying across the galaxy's center. But others have speculated that this ring was expelled by a titanic explosion that occurred in the heart of the Milky Way a few tens of millions of years ago.

What might have been the nature of such an explosion? No one really knows at this time, but it could be closely linked to another energetic event that seems to have taken place around the same time. Using sensitive infrared telescopes, University of Arizona astronomers Marcia

Lebofsky and George Rieke, among others, have detected a large collection of red supergiant stars in the galactic center that look as if they had formed in a burst of star formation only ten or more million years ago (by astronomers' notion of time, that's practically yesterday—several "days" ago in the lifetime of our Sun). Lebofsky and Rieke estimate that, during this burst, the galactic center was converting gas into stars a thousand times faster than any other place in the galaxy. This might have been an isolated incident, or it may be an indication that our galaxy's core somehow undergoes periodic spurts of stellar production every few tens of millions of years. "I remember reading in the textbooks of the 1950s that the galaxy's core was a rather quiet place where stars sat out their retirement awaiting a slow death," says Rieke. "Now we know that's not true at all. Starbursting is not an uncommon phenomenon."

To add to the mystery, astronomers have encountered a few more "smoke rings," complicated mixtures of molecular clouds and energized gases, as they probe farther and farther inward to the galaxy's heart.

The innermost sanctum of the nucleus, a region about a dozen light-years in diameter, is the most densely packed section of the galaxy. Millions of stars reside there. By comparison, a 12-light-year-wide region out here in the fringes of the disk contains only a couple of handfuls of stars. We sit in a virtual tundra, while the center resembles a teeming metropolis. Infrared telescopes, which have the fortunate ability to pierce the center's dusty curtain, are able to discern a cluster of stars dwelling there, looking exactly like phosphorescent gems set against a velvety black curtain. Gareth Wynn-Williams of the University of Hawaii once fancifully described what it might be like for an astronomer to live fairly close to this star-filled hub: "The brightest single object would be a brilliant red star . . . some 100 times brighter than Venus appears. . . . In most directions our astronomer would see many more bright stars than he would see if he were on Earth, and the total light from them would exceed full moonlight [200 times over].

"The majority of the stars would be to one side of him, in the direction of the nucleus of the Galaxy, in a giant cluster increasing in brightness to its center. From this center would emanate radio waves 10,000 times more intense than any signals detected by radio astronomers on Earth. All around the astronomer would be pink and green clouds of ionized gas illuminated by hot blue stars. As they orbit the nucleus of the Galaxy, many of the stars would change their apparent positions in the sky far more rapidly than do the Sun's neighbors, so that in the astronomer's lifetime the visual appearance of the sky would change significantly."

Unfortunately, such a magnificent spectacle would not be without its drawbacks. The dust and gas in that region, combined with interference from the appreciably bright "nighttime" sky, would essentially block any decent view of other, far-off galaxies. It is interesting to imagine the type of cosmology that alien astronomers from the galactic hub might concoct. In all likelihood, they would develop a model of the universe vastly different from ours.

As astronomers aim their telescopes directly at the galaxy's heart, the terrain becomes a virtual jungle of radio, infrared, and gamma-ray data difficult to decipher. They can see that a doughnutlike ring of gas and dust, some twelve light-years wide, surrounds the center, and that the region inside this turbulent hoop of silicate grains and molecular gas is appreciably scoured of celestial debris (aside from the stars themselves). Did the intense radiations from a particularly dense concentration of stars vaporize the dust in that sector, or was it blown away by some other, unknown cataclysm? Some clues are provided by measuring the temperature to which the dust ring is heated; in total, it contains the equivalent energy of tens of millions of our Suns. Could stars alone, shining and exploding nearby, be the originating source of all that heat? Possibly. Such a mechanism cannot, as yet, be ruled out. But some astronomers argue that it would require a type of cluster whose existence is hard to imagine. As a result, these astronomers contend that a more unusual object must be sitting at the very center of the dust ring, dominating the stellar scene like the bull's-eye in a dart board. The most popular candidate is a black hole. This idea was first advanced by British astrophysicists Donald Lynden-Bell and Martin Rees as early as 1971.

Over the last decade, evidence has been slowly building to back up this suspicion. Early on, astronomers from the University of California at Berkeley detected small clouds of ionized gas swirling around the center at very high velocities. To keep this gas from flying away, it has been suggested that the clouds are satellites of a supermassive object hiding out in the galactic core. The dynamics of the situation suggest that this "compact source," as astronomers like to call it, contains the mass of a few million Suns. Theorists speculate that if it is truly a black hole, some of the prolific radiations emanating from the galactic nucleus are generated as the hole devours anywhere from 1/1,000 to 1/100,000 of a solar mass a year, converting part of the material into radiant energy before the matter is irretrievably lost down the bottomless space-time well. There is certainly an ample food supply in the area. Since a few million stars reside in the central few light-years, ejecta from stellar winds, supernova explosions, and planetary nebulae could easily feed the elusive "beast."

Astronomers would have been terribly frustrated in their efforts to resolve our galaxy's central powerhouse in greater detail, if it were not for the construction of radio astronomy's most powerful instrument, the Very Large Array. Since it is not feasible to construct an immense radio telescope many miles in diameter, astronomers took advantage of advances in computer technology to make a set of individual radio dishes mimic the capability of such a single miles-wide antenna. By letting the Earth's spin sweep the antennas around, an astronomer, with the aid of a computer to keep track of the movement, can build up an immense imaginary radio telescope segment by segment. Following this strategy, the 27 dishes in the VLA system can sweep out a receiving area as much as 21 miles across. Because this synthesized dish has a diameter so much larger than the individual dishes, it can resolve much finer details in any given radio source. Every second, computers located in the VLA's main control building digest a million bits of data arriving from the array via a 40-mile-long system of waveguides, two-inch copper-lined steel tubes buried beneath the desert plain. A typical 12-hour observation can take up three computer tapes, enough digital information to fill some two hundred books.

Each datum helps define the intensity of the radio source at a certain point in the sky. Once these myriad bits are processed and displayed on a computer graphics terminal, a color-coded radio picture of the source emerges. One particular strength of the VLA is its ability to act like a giant zoom lens. A few months at a time, the antennas are crowded in, each arm of the *Y* no more than half a mile long; this provides a sort of wide-angle view, perhaps to trace the gas clouds in a nearby galaxy. But to get a closer look, the antennas are periodically moved along railroad tracks out to greater distances, up to 13 miles along any one arm. To a source in the sky, this configuration appears to fashion a dish larger than Washington, D.C., as the Earth slowly turns on its axis.

When astronomers, among them Ronald Ekers and Robert Brown of the National Radio Astronomy Observatory, first employed the VLA to home in on the galactic center, they discovered that the energized gas swirling around the hub actually forms a spiral-like pattern, a few light-years wide, which is centered on an extremely bright and extremely tiny radio source. The VLA computer-generated picture looks deceptively like a multiarmed spiral galaxy. At first, it was thought the gaseous arms were some sort of jets spewing from the compact source, like a larger version of SS 433. But a few years later, Caltech astronomer Lo and his associate Mark Claussen, after making a more detailed map with the VLA, concluded that the spiraling streams were more likely being drawn *into* the center. It's still quite plausible that all this material is being pulled inward gravitationally by an extremely dense cluster of stars perched at the galaxy's hub. But Lo believes that his VLA observations strengthen the case for a single supermassive object. If such a model of the galactic center is correct, it would be the first time that astronomers are directly seeing matter fall into a black hole.

Using an intercontinental array of radio telescopes, Lo and several associates were able to set an upper limit on the size of the compact radio source sitting smack-dab in the middle of the spiraling gas: The central object appears to have a diameter no greater than the width of Saturn's orbit (about two billion miles), maybe less, a dimension that makes the supermassive-black-hole hypothesis even more compelling, though still not conclusive.

Over time, a multitude of instruments, including x-ray telescopes and gamma-ray detectors, have been aimed at the Milky Way's core. Gamma rays from the galactic center were discovered as early as 1970, by a detector carried aboard a high-altitude balloon launched over the outback of Australia (a propitious site to study the Milky Way's core, since it is best seen from southern latitudes). The gamma-ray signals intermittently emanating from that direction are unmistakable: At least half the rays are at one specific energy, the same energy released whenever an electron slams into its antimatter mate, the positron, annihilating both elementary particles in a hail of gamma radiation. And, again, the black-hole model comes into play. Theorists claim that such atomic particles would most likely be created within or near an accretion disk of hot plasma—a disk that would inevitably encircle a black hole.

If a black hole does power at least part of the tremendous activity taking place in the galactic center, still it is weak by cosmic standards. There are galaxies in our universe that emit hundreds, even thousands of times more energy from their cores. Despite the violent turmoil described above, the nucleus of the Milky Way could be considered as experiencing a relatively "quiet" phase at this moment, possibly caught between bouts of more intense outpourings. Speculation abounds on how a black hole might be associated with all the varied energetics observed in the nucleus. Perhaps the assorted rings of material situated right near the center were produced when the "beast" got hungry and ate a little bit more than usual, somehow triggering both bursts of star formation and battering shock waves. The innermost ring surrounding the galaxy's center may have been swept out by the last shock wave, and the spiraling streams could be clumps of gas that are starting to fall back in, readying the hole for its next explosive outburst. The story emerging from our galaxy's center is far from over.

The latest chapter being written by the electromagnetic radiations escaping from the galactic center discloses that the Milky Way could also be acting like a giant electric dynamo. The discovery came about very innocently. A team of astronomers from Columbia University and the University of California at Los Angeles were using the Very Large Array to search for definitive signs of star formation in the center of the galaxy. But instead of detecting the telltale symptoms of stellar birth, they noticed, to their astonishment, an enormous arc of hot, radio-emitting gas spewing from the galaxy's heart. Composed of several parallel strands, the hook-shaped feature was vaguely discerned in the 1950s, but never imaged so crisply or clearly as with the VLA. "In the chaos of the interstellar medium," notes UCLA astronomer Mark Morris, one of the arc's discoverers, "such regularity is quite abnormal." It is so unusual, in fact, that Morris and his Columbia colleagues Farhad Yusef-Zadeh and Don Chance were reluctant to accept the finding for several weeks, thinking it might be the result of an error in the data processing.

To keep the tight bundle of threads so perfectly aligned over distances of 150 light-years, astronomers postulate that the glowing arc is stretched along magnetic-field lines that run perpendicular to the plane of the galaxy, similar to the field lines jutting out of Earth's and Jupiter's polar regions. Just as the motions of a fluid iron core are thought to generate the Earth's magnetic field, so too might a "poloidal" magnetic field be produced as electrically charged gases in the central regions of the disk perpetually move around the nucleus.

To add to the intrigue, the glowing filaments, so beautifully imaged by the VLA, may be only one segment of a much larger structure that was "photographed" by another team of radio astronomers at just about the same time as the arc was discovered. Using a 150-foot-wide dishlike antenna at the University of Tokyo's Nobeyama Radio Observatory, Japanese observers detected a horseshoe-shaped plume of radio emission majestically rising some 700 light-years above the galactic plane. The smaller filamentary arc sits at the bottom, at one end of the horseshoe. Astronomers can't help but notice that the monstrous plume closely resembles the giant loops of scorching plasma often seen bubbling away from the Sun's surface. Though the galactic version is billions of times bigger, it's very possible that the mechanism producing the galactic loop of fiery gas is not too different from the solar kind. Magnetic lines of force, coiled within the plane of the disk like cable wire wrapped around a spool, could be getting wound up like a watch spring as the galaxy rotates. Tension in this galactic spring could then build up to the bursting point, whereby the field lines "pop out" and enable radio-wave-emitting electrons to stream out along a new magnetic pathway. On the other hand, it might be some kind of exhaust jettisoned from the galaxy's active nuclear region.

Whatever the mechanism, the highly organized arc and plume have taught astronomers one unquestionable lesson: that magnetism—the same force that lines up iron filings around a bar magnet—could be playing as large a role as gravity in sculpting some very prominent and large-scale features in the Milky Way galaxy. It is a trick that nature often pulls on astronomers. New data can lead, not to firm answers, but rather to new questions and new ways of perceiving the motor that drives and shapes the universe's many creations.

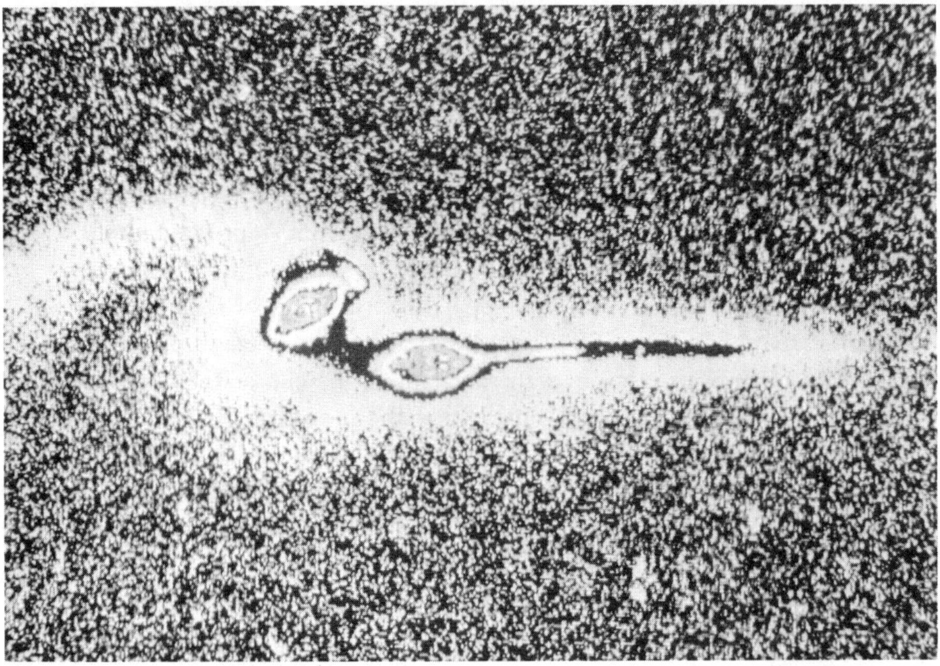

At one time galaxies were thought to be immutable islands of stars, tranquil and isolated. Now it is known that they can be drastically altered through interactions with their neighbors. Above, two galaxies (dubbed the Playing Mice) collide, irrevocably changing their structures. *Courtesy of National Optical Astronomy Observatories.*

5

THE COSMIC MARATHON

I see the lines of your spectrum shifting red,
The universe expanding, thinning out,
Our worlds flying, oh flying, fast apart.

—Stanley Kunitz, *The Science of the Night*

For countless decades, the Milky Way was, by default, the sole galaxy in which astronomers could observe and gauge celestial behavior. The naked eye and, later, crude telescopes could focus only on the near shores of space, not the far horizons. The universe and the Milky Way, with its enveloping shroud of flickering lights, were one.

Astronomers might have developed a complacent attitude toward this state of affairs, never considering further hierarchies in the universe's structure, were it not for certain wispy patches of light in the nighttime sky that could not be resolved into individual stars with either eighteenth- or nineteenth-century telescopes. Moreover, unlike such gaseous clouds as the Orion nebula, these tiny fuzzy objects tended to be located away from the plane of the Milky Way. William Herschel sighted thousands of them through his telescope. Though the varied hypotheses concerning these faint, peculiar formations were at first impossible to prove, the mysterious objects turned out to be beacons leading astronomers to ask important new questions. And the answers were revolutionary.

The most prominent of the fuzzy formations, the Great Nebula in the Andromeda constellation, was noted on star charts centuries before the invention of the telescope. The German astronomer Simon Marius, a contemporary of Galileo, wrote that the long oval-shaped spot bore a resemblance to "the light of a candle which one sees from a distance in the night through a piece of transparent horn." By the middle of the nineteenth century, the earl of Rosse, William Parsons, electrified the astronomical community by announcing that many of the white nebulous patches in question exhibited a distinct spiral structure. Since photography had not yet been introduced to astronomy, Lord Rosse captured the swirling patterns in drawings of splendid detail with which he filled his notebooks. Many look exactly like the gyrating hurricanes that weather-satellite photos every so often show meandering across the Earth's oceans. These spiral-like nebulosities lurked in the background of space, pulling at astronomy's shirtsleeves like pesky children. What were they?

By the start of the twentieth century, astronomers were essentially divided into two camps over this question. After simmering on the back burner for quite a while, the nagging controversy was given a public airing on April 26, 1920, when the leading proponent for each side, Heber Curtis of California's Lick Observatory and Harlow Shapley of Mount Wilson, squared off at the National Academy of Sciences in Washington, D.C. Although Shapley later remembered it as more of a symposium,* historians have since labeled this showdown "The Great Debate."

During his turn at the podium, Curtis contended that the spiral nebulae were independent stellar systems situated well outside of the Milky Way. "The spirals are not intragalactic objects but island uni-

*The advertised subject of the debate was actually the size of the Milky Way. Shapley had recently introduced his controversial new model of the galaxy. But the discussion of a crucial side issue—the nature of the spiral nebulae—is best remembered.

verses like our own galaxy," he argued. This idea, though controversial at the time, had not originated with Curtis and his supporters. The term "island universe" is usually attributed to Immanuel Kant. As early as the 1700s, both Thomas Wright and Kant had speculated that some of the mysterious nebulosities sighted by observers might be far-off cousins of the Milky Way. "Let us imagine a system of stars gathered together in a common plane, like those of the Milky Way," wrote Kant in 1755, "but situated so far away from us that even with the telescope we cannot distinguish the stars composing it. . . . Such a stellar world will appear to the observer, who contemplates it at so enormous a distance, only as a little spot feebly illumined. . . ."

In Kant's day, such a view, though surely farsighted, was pure conjecture. But by 1920, Curtis had some tentative evidence to support this contention. A few years before the debate, George Ritchey of the Mount Wilson Observatory had discovered, entirely by chance, a brilliant starlike object, presumably a nova, in a photograph of the diffuse spiraling cloud known as NGC 6946 (from John Dreyer's *New General Catalogue* of nebulae and star clusters published in 1890). The stellar explosion was not evident on plates made earlier. By examining other photographic plates at Mount Wilson, Ritchey saw that two novae had also appeared in the Andromeda nebula. Spurred by the announcement, Curtis went through the Lick files and spotted novae in two other spiral nebulae. The apparent faintness of some of the erupting stars, when compared with the luminosity of nova outbursts in our own galaxy, implied that the spiral nebulae were stellar systems situated far beyond the outer borders of the Milky Way—at least by Curtis's calculations.

Shapley strongly disagreed. In hindsight, it appears as if Shapley was being obstinate, but the evidence was truly muddled. For example, two of the outbursts cited by Curtis were particularly luminous; if they were ordinary novae, it could be argued that they were not more than a few thousand light-years away, well within the boundaries of the Milky Way. But astronomy did not yet recognize the existence of supernovae, which ferociously blaze with the intensity of a billion suns. Furthermore, many observers were still skeptical of claims that stars had actually been sighted in the spiral nebulae; to their trained eyes, a spiral's light spectrum looked strangely different from what was then known about stellar spectra. And, since an interstellar medium of dust and gas had not yet been detected and taken into account when making distance measurements, Shapley continued to believe that the disk of the Milky Way had enormous dimensions, making it possible for the spirals to be merely minor entities caught at the edge of one huge system. A few scientists even wondered whether the spirals might be globular clusters or planetary systems in the making.

Shapley's strongest argument against the island-universe proposition (to his later dismay) was based on completely erroneous data. Shapley's friend and colleague at Mount Wilson, Adrian van Maanen, had measured a rather high rate of rotation in certain spirals, which indicated that the nebulae had to be close by, inside the Milky Way. If they were particularly far away, it was reasoned, such rapid spins could not be ascertained unless the stars and gas were rotating faster than the speed of light—an impossibility. Van Maanen's measurements were later proven to be incorrect, but at the time his finding was often used as powerful testimony against the Curtis school of thought.

There was no real winner in this famous debate. Forty-one years later, in 1961, Allan Sandage, the eminent American astronomer, diplomatically concluded that "Shapley used many of the correct arguments but came to the wrong conclusion. Curtis, whose intuition was better in this case, gave rather weak and sometimes incorrect arguments from the facts, but reached the correct conclusion."

Like so many disagreements in astronomy, the issue was resolved with the development and use of more powerful instruments. Just a few years after the massive Hooker telescope, with its awesome 100-inch light-gathering mirror, was assembled on the peak of California's Mount Wilson, astronomers entered the age of extragalactic studies, the examination of systems far beyond the Milky Way. It was a portent of the dramatic changes in astronomical concepts that would occur throughout the twentieth century, whenever astronomy either greatly improved its optical instrumentation or explored new regions of the electromagnetic spectrum. The first vital step in this extragalactic research was taken by a Renaissance man and showman named Edwin Hubble.

Between 1906 and 1913 Edwin Hubble was an accomplished boxer during his undergraduate days at the University of Chicago, turning down a chance to turn professional; a student assistant in the laboratory of the great American physicist Robert Millikan; and a Rhodes scholar in Roman and English law at Oxford University. But astronomy was his passion from his earliest days. A year after his return from studying law in England, Hubble decided to "chuck the law for astronomy" and returned to the University of Chicago to enroll in its graduate astronomy program. He said he was making this career change even if it meant being a second-rater. But the ex-law-scholar perceptively chose to undertake a doctoral thesis on one of the most important topics of the day: the nature of those enigmatic nebulae. This decision to enter into the "realm of the nebulae," as he described it, placed Hubble at the forefront of his newly chosen field. And, like the experts before him, he concluded that answers to his questions must await "instruments more powerful than those we now possess."

He did not have too long a wait. In 1919, after a two-year hitch in the army, where he attained the rank of major while fighting in France at the end of World War I, Hubble went on staff at the Mount Wilson Observatory. By then, the long-awaited 100-inch telescope was beginning to resolve individual stars on the outskirts of some of the spiral nebulae. Patiently working with the giant reflector over several years, Hubble was able to identify, in particular, a number of Cepheid variable stars in Andromeda and two other nebulae. It was a terrifically exciting find; these were the same stellar beacons that had enabled Shapley to size up the Milky Way. The period of the Cepheids' blinking, together with the extreme faintness of their magnitudes, enabled Hubble to deduce that the nebulae were located very far out in space, in a region beyond the limits of even Shapley's overlarge model of the Milky Way.

Hubble calculated that the Andromeda nebula, for one, was nearly 800,000 light-years away from us. This distance was later adjusted upward to two million light-years (once Walter Baade recognized in the 1950s that there are different types of Cepheid variables with different period-luminosity relationships that must be used to mark distances), but Hubble's initial estimate still placed the Andromeda nebula drastically farther away than astronomers ever dared hope to find celestial objects. In one stroke, he had redefined the boundaries of the universe. The "Great Debate" was speedily resolved.

Hubble spoke of our remote neighbors as "extragalactic nebulae," yet in time astronomers came to refer to the distant stellar systems as "galaxies," recognizing the fact that the Greeks' original *galaxias kyklos* or milky circle, our own Milky Way Galaxy, was just one of many systems roaming through the seemingly endless gulfs of space. Our solar system is a mere mote in their presence.

Until his death in 1953, Hubble devoted himself to studying these newfound members of the universe. Like a botanist sorting through a variety of flora, Hubble became an extragalactic taxonomist. He sought out the various species, described their properties, and divided them into major classes. Hubble fairly quickly recognized two broad categories of galaxies: spirals and ellipticals. Of the hundreds of prominent galaxies Hubble photographed with the 100-inch telescope, two-thirds were found to be of the spiral type, containing anywhere from ten billion to nearly a trillion stars. The Milky Way is an average-sized specimen. Every spiral galaxy is essentially a gigantic stellar pinwheel: a bright central bulge surrounded by a thin disk of stars and interstellar matter. Parts of the disk stand out as luminous spiral arms. In some cases, the arms are wound tightly around the bulge; in others, the limbs are spread wide. In barred spirals, the arms form at the ends of a barlike structure. Although old red stars mostly comprise the bulge, the overall light from a spiral galaxy tends to be bluish because hot and massive young stars continue to form in the arms.

An elliptical, on the other hand, is a much denser galaxy, shaped somewhat like an egg, that appears more reddish than a spiral because most of its stars are very old. This type of galaxy has largely exhausted its supply of dust and cool gas, the raw materials for new star formation. You can think of an elliptical galaxy as a large central bulge without the accompanying disk. These stellar "footballs" have a great range of sizes. The so-called "dwarf" ellipticals contain only a million or so solar masses (a hundred-thousandth the mass of the Milky Way). The very largest ellipticals, though, can harbor the suns of a hundred Milky Ways.

A type of galaxy that falls in between a spiral and elliptical is classified as an S0 galaxy, which consists of a sizable bulge enwreathed with a completely smooth disk of stars. It is devoid of spiral arms. An S0 galaxy is akin to an elliptical in that it has little gas with which to make new stars.

Hubble also recognized that a certain percentage of galaxies are rather loose aggregations. Such "irregular" galaxies are chaotic clumps of stars immersed in a very rich sea of gas. Examples of this form are the two small associations tagging alongside the Milky Way: the Magellanic Clouds, which have been dubbed the "crown jewels of the southern skies." While spiral galaxies, as a class, hoard the largest portion of luminous matter in the cosmos, dwarf irregulars, along with dwarf ellipticals, are actually the universe's most numerous inhabitants.

It is estimated that some 100 billion galaxies, big and small, reside in the observable universe, extending in all directions as far as we can telescopically see. The intricate lenses and mirrors of the world's largest telescopes have revealed that there are as many galaxies in the universe as there are stars in the Milky Way.

The administrative offices of the Mount Wilson (California) and Las Campanas (Chile) observatories are set amid the decorous homes and tree-lined streets of Pasadena, where, decades ago, orange groves basked in the California sun. Astronomers from around the globe simply call the complex "Santa Barbara Street." In his office at 813 Santa Barbara Street, astronomer Alan Dressler pulls a large 20-inch by 20-inch glass plate out of a cardboard container and carefully props it up before a light box to peruse a dense cluster of galaxies known as Abell 548 (from George Abell's catalog of particularly rich groupings of galaxies). The thick plate is a photographic negative; details can be seen more clearly when the sky is light and the celestial objects are dark.

What immediately strikes the uninitiated is the enormous number of galaxies dotting the plate: not one or two, but literally dozens of faint, fuzzy smears no bigger than the tip of one's finger. It is a universe in miniature. And this picture, taken with the 100-inch telescope at the Las Campanas Observatory situated in the foothills of the Chilean Andes, covers only 1/20,000 of the celestial sphere. Some of the spots on the plate can be identified as featureless ellipticals; others are unmistakably resplendent spirals with star-forming regions standing out like glowing bubbles along the arms. For someone with an Earth-centered perspective, this cornucopia of galaxies is an unsettling sight. From such a standpoint, a galaxy becomes just a basic cell of the cosmos. But whereas a biological cell converts raw materials into energy to nourish our body, a galaxy in essence transforms hydrogen gas into heavier elements through the creation of stars. Planets, in many respects, are a very minor sideline.

What is not immediately evident in the picture of Abell 548 is the intrinsic dynamism of the universe. Galaxies are in flight. With each second, the time it takes to flick a light switch, the ellipticals and spirals in Abell 548 rush 10,000 miles farther away from the Milky Way. In an hour, they recede 36 million miles. Vesto Slipher detected this exodus as early as 1912, before the spiral nebulae were even unmasked as individual galaxies. He did this by studying their spectra (the composite wavelengths of light or colors emitted by the nebulae). Working for many years at the Lowell Observatory in northern Arizona (a facility built by the wealthy Percival Lowell to study the "canals" of Mars), Slipher obtained the spectra of several dozen nebulae. Spectroscopy at that time was a painstaking task; many hours, often successive nights at the telescope, were required to obtain the spectrum of just one nebula. In the vast majority of his cases, Slipher found that certain spectral features, well-known emissions of select elements, had shifted toward the red end of the electromagnetic spectrum. Such a change was a telling clue.

These "redshifts" told Slipher that the spiraling clouds were rushing away from Earth at fairly high velocities. A comparable phenomenon occurs on the highway every day. We've all heard the pitch of a siren rise particularly high as an ambulance races toward us. This is the well-known Doppler effect: The sound waves emitted by the screeching siren crowd together as they approach us, shortening the length of the sound waves and likewise raising the pitch detected by our ears. Conversely, as the ambulance recedes, the sound waves stretch out, producing a lower pitch. In an analogous fashion, a light wave's length is shortened (gets "bluer") when the source of the light approaches and is lengthened (gets "redder") when the source recedes. The higher the speed at which a celestial object rushes either toward or away from us, the greater the shift in its light; blueshifts and redshifts are the speedometers of the universe.

When spiral nebulae were still thought to be members of the Milky Way, Shapley speculated that our galaxy could be exerting a peculiar force of repulsion against the spirals. But astronomers soon learned that the fleeing nebulae were actually separate galaxies. By 1929, Hubble had determined the distances to about two dozen galaxies. Coupling this distance information with data on the galaxies' redshifts, he was able to find order in the cosmic marathon first sighted by Slipher: The more distant a galaxy, Hubble saw from the data, the faster it moves away from us. This was one of the most profound discoveries in twentieth-century astronomy. The conclusion was inescapable. The entire universe visible to our eyes was expanding. The spark that ignited this tremendous push outward, it is now generally assumed, originated in a "Big Bang"—the cataclysmic explosion that is believed to have given birth to our universe many eons ago (see chapter 9).

Teamed with Milton Humason, Mount Wilson's crackerjack redshift measurer, Hubble went on to investigate more than a hundred ever dimmer, more distant galaxies. The results only confirmed Hubble's original announcement: A galaxy's redshift was directly proportional to its distance from the Earth. Space-time is steadily expanding, and galaxies are going along for the ride. Aptly enough, the parameter that describes this rate of expansion has come to be known as the Hubble constant, a term which is commonly expressed in units of kilometers per second per megaparsec.*

In the thirties, Hubble calculated that the rate of expansion stood at around 530 kilometers per second per megaparsec (3.26 million light-years). That meant that any two galaxies separated by one megaparsec would be speeding away from one another at 530 kilometers (330 miles) per second because of the universe's expansion. Likewise, galaxies situated two megaparsecs apart would be receding at 1,060 kilometers per second. Doubling the distance doubles the velocity, tripling the distance triples the velocity, and so on down the line.†

This concept is not easy to grasp. Even an expert cosmologist like Princeton's P. James E. Peebles has said that "it's anti-intuitive and against all experience." Since virtually all galaxies appear to be running

*A parsec is an odd astronomical unit of distance equal to 3.26 light-years. It is the distance at which a star exhibits a parallax of one second of arc, an apparent shifting of its position on the sky, due to the Earth's motion around the Sun.

†Virtually the only galaxies not racing away from us are the Milky Way's closest neighbors, which are bound to one another through gravity and so travel through the universe together. Andromeda's spectrum exhibits a blueshift that indicates it is moving toward us at nearly 200 miles a second.

away from us, one gets the impression that *we* are poised at the center of both the universe and the original explosion, but this is not the case at all. Hubble discovered that, in a way, galaxies are like raisins in an unbounded loaf of raisin bread baking in a cosmic-sized oven (a popular analogy used by many astronomers). As the dough (space-time itself) swells outward in all directions at a certain rate, each raisin set in the dough sees every other raisin recede from itself. And, within such a ballooning loaf of bread, raisins set far apart simply separate faster than raisins located near one another, which is why a raisin situated twice as far away as another appears to recede twice as fast. From the viewpoint of any single raisin, all the other raisins are moving away from it in a proportional manner. In this way, every raisin (galaxy) thinks it is the unmoving center of the dough (universe).

Working the expansion backward in time also gives astronomers a rough handle on the age of the universe. Imagine being able to shift a magical cosmic gear that puts the swelling of space-time into reverse. The higher the Hubble rate, in some sense, the less the time needed to get back to the Big Bang and the smaller the observable universe. With a Hubble constant of 530, the universe turned out to be 1.8 billion years old, an awkward result once geologists were positive that the Earth was more than *4* billion years old. A great sigh of relief could be heard among astronomers during the 1950s and '60s when vastly improved distance and velocity measurements by both Baade and Sandage, Hubble's protégé at Mount Wilson, eventually brought the Hubble constant down to a sedate 50 kilometers per second per megaparsec. This corresponds to a universe nearly 20 billion years old—time enough to create all the galaxies, stars, and planets. Hubble, it turns out, had mistaken other kinds of objects, such as nebulae, for stars when he measured the distances to more and more distant galaxies, and Sandage, for one, was able to correct for these errors with the higher sensitivity of the 200-inch telescope on Palomar mountain.

Actually, it is ironic that the Hubble parameter has come to be labeled a "constant," for it has turned out to be a rather mercurial number. By the 1970s, astronomers such as Gerard de Vaucouleurs of the University of Texas at Austin and Marc Aaronson of the University of Arizona's Steward Observatory, working with Jeremy Mould of Caltech, completed new distance measurements, the linchpin in any Hubble-constant determination, that suggested the rate of expansion is more like 85 or 95 kilometers per second per megaparsec. Some interpret this higher figure as meaning that the observable universe is both much smaller and much younger—perhaps, a mere 10 eons old—than was originally thought. But, as pointed out in the last chapter, this creates a paradoxical situation: The stars in globular clusters appear to have formed some 15 to 18 billion years ago. Another cosmic clock, the ages of elements as gauged by the rate of their radioactive decay, seems to agree with the globular clusters.

"Astronomers like to ask very simple questions, such as, 'How big is the universe? How old?' But, surprisingly," says Aaronson, "these simple questions cannot as yet be answered with certainty." One problem lies in the fact that each group attempting to determine the Hubble constant makes different assumptions in analyzing their data. Sandage and his Swiss collaborator Gustav Tammann, for instance, differ from de Vaucouleurs in the way they estimate the amount of obscuring interstellar dust in our galaxy that lies along the line of sight between Earth and the galaxies, and this affects the extent to which one's chosen "standard candle" decreases in brightness with distance.

And what kind of standard candle is best? Everyone has a favorite. Pulsating Cepheids are visible only in the nearer galaxies. To reach beyond that, astronomers must search distant galaxies for a host of other range indicators: novae and supernovae, particularly bright stars, globular clusters, and glowing clouds of ionized hydrogen similar to the Orion nebula. Aaronson and his associates make their distance measurements by first determining the rotation speed of a spiral galaxy (which is linked to the galaxy's inherent luminosity) and then ascertaining the apparent luminosity of the spiral in the infrared, since infrared wavelengths are less affected by intervening dust. Attempts have also been made to define a "standard ruler," such as the diameter of a certain type of galaxy or nebula, and then gauge how such objects appear to diminish in size with distance. Complicating the issue, though, is the fact that distance measurements in the universe involve a shaky series of stepwise calculations, each rung of the ladder dependent on the one below it, and key calibrations are often disputed among competitors in this field.

Distance determinations begin with the stars closest to the Sun: Their distance is measured trigonometrically by noting how much they change position against the celestial background when they are observed first from one point in the Earth's orbit around the Sun and then, six months later, from the other side of the orbit. Using those results as a calibration, the distance of stars and galaxies farther out is estimated by a complex sequence of computations. One mistake in the chain can throw off measurements to the edge of the universe.

Unfortunately, rather than forging a consensus, all the varied efforts have led to Hubble "constants" ranging between 100 and 50.* That roughly translates into a universe that is anywhere from 10 to 20 billion years old. One cynic resolved the problem by suggesting that the universe is twice as big on Mondays, Wednesdays, and Fridays as it is on Tuesdays, Thursdays, and Saturdays. On Sundays, he said, it's a toss-up. For the moment, there is no one method that can serve as the tie breaker in this deadlock. If the higher Hubble-constant value is upheld, astronomers must still face the dilemma such a revision presents: 15-billion-year-old globular clusters in a 10-billion-year-old universe. If that turns out to be the situation, it's quite possible that Einstein's equations of general relativity, long used to describe our cosmic expansion, will have to be revised slightly. Theorists may be forced to reintroduce Einstein's infamous cosmological constant, which the great scientist himself described as the biggest scientific mistake of his life.

Out of desperation, Einstein tacked the extra term onto his equations in 1917. His original, unaltered theory of general relativity posited that the cosmos was dynamic. But then-current observations of the heavens suggested that the universe was static and unchanging, prompting Einstein to formulate the cosmological constant to resolve the impasse. The term, as he used it, said there was a repulsive force at work in the universe, a kind of antigravity that exactly balanced the gravitational attraction of the galaxies, keeping them from moving. But once Hubble revealed that our universe was truly expanding, Einstein quickly (and gladly) dropped the term. He always believed it sullied the symmetric beauty of his gravitational equations.

"Physicists often say that Einstein's equations are more aesthetically pleasing without the constant," says de Vaucouleurs, "but that's not a

*Throughout this book, extragalactic distances and ages are given assuming a compromising Hubble constant of 75.

good reason to dismiss it. We like things to be simple in science, but nature usually turns out to be more complicated." For those in favor of a higher Hubble rate, such as de Vaucouleurs, inserting a small cosmological constant back into Einstein's equations easily resolves the problem of the universe's age. The repulsive push inherent in the constant dictates that the universe would have expanded more quickly since the Big Bang, thus making the cosmos a bit older than simple Hubble calculations suggest; it can even produce a universe as old as 18 billion years, a pleasing result for those who are worried about aged globular clusters. Of course, many astronomers are still convinced that the Hubble constant is closer to 50, making all such extra manipulations unnecessary.

"Considering the immensity of the job," says University of Washington astronomer Paul Hodge, who has been involved in calibrating the cosmic distance scale, "we should feel fortunate that we are now able to argue about only a factor-of-two uncertainty. . . . Rather than be disturbed and discouraged that the scale of distances is still a matter of dispute, we should probably be grateful that nature has made it possible for us to find methods by which we can measure something so incredibly big and so immensely important as the universe." Just as the mystery of the spiral nebulae was quickly solved through the use of Mount Wilson's 100-inch telescope, it is widely hoped that this spirited debate over the universe's size and age will be settled once the next generation of space and ground-based telescopes, currently under construction, begin gathering their first streams of celestial light.

As soon as the notion of an expanding universe was firmly rooted in astronomers' psyche, the next question for cosmologists was an obvious one: Is the universe open and will it thus expand forever—like a rocket that escapes from the Earth's gravitational pull, never to return —or will it, at some far, far moment in the future, slow to a halt and then close back up, pulled inward by the combined gravity of all its mass as if the galaxies were tethered by invisible rubber bands? This is considered the most crucial unanswered question still facing cosmologists. In Hubble's day, distant galaxies were studied largely in hope of answering this riddle: open or closed? Astronomers' prime motivation in tracking galaxies at all was their insatiable desire to learn of the universe's fate. "Galaxies were just little markers, spots in space, to enable astronomers to discern the geometry of the universe," notes Princeton astrophysicist Jeremiah Ostriker. This was (and still is) their grandest quest.

Should the universe start to collapse many eons from now, our future is assured: Over billions of years, the universe will get squeezed more and more—first galaxies, then stars and planets, pressed into one another—until the cosmos ultimately reforms the dense and brilliant fireball of its youth. But if expansion continues forever, a terribly bleak destiny will surely prevail. Eventually depleted of their supply of gas, galaxies will stop making stars, and the stars already in existence will, in time, flicker out. Red dwarfs and white dwarfs, the major stellar populations, will cool to black cinders. Galaxies themselves will become extinct. In a billion billion years or so, the dense cores of galaxies will likely turn into mammoth black holes, comprised of the mass of a billion suns. The rest of the galaxy—its leftover gas, dust, asteroids, planets, black dwarfs, neutron stars, and stellar black holes—will gain enough velocity from countless stellar encounters to escape from the galaxy altogether and wander off into space, destined to disintegrate completely if matter decays. The universe will become an ever dimmer and duller place to live. It will revive slightly around 10^{67} years from now when black holes, which have long been emitting elementary particles, start disappearing altogether in bursts of gamma rays. But by 10^{100} A.D., even these occasional fireworks will be gone forever, with the released energy condemned to join a vastly diffuse and cold sea of radiation.

For many years, it was widely believed that astronomers would be able to determine which of the two fates will befall the universe. The tests were deemed difficult but feasible. The fact that light waves travel at a finite speed was the key. Since the light from a star or galaxy is not instantaneously transmitted to Earth, we are constantly viewing the universe as it "was," not as it "is." Even the Sun's rays take a full eight minutes to travel the 92 million miles to Earth. The familiar stellar disk we see daily in the sky is always eight minutes old. Likewise, the light waves emitted by far-off galaxies take millions to billions of years to arrive at our telescopes. So, the farther we peer out into the universe, the further we are looking back into its history. In principle, that means we should be able to observe an era in the remote past when the overall expansion of the universe was proceeding a little faster than it is now, since the earlier epoch would be closer in time to the explosive Big Bang. By comparing that past speed with today's rate of expansion, astronomers could conceivably judge whether galaxies are slowing down enough to someday stop in their tracks and eventually fall back in a "Big Crunch," or are being carried outward at an unstoppable speed.

Sandage has been a leader in performing these cosmological tests. The brightest galaxy in a cluster—its average size, shape, and luminosity—was often used as his cosmological yardstick. It was a viable effort, as long as galaxies could be thought of as changeless markers, objects that have drifted on in tranquil isolation experiencing little or no evolution after their tumultuous birth billions of years ago. Using a galaxy as a yardstick is fine—if the galaxy stays the same size through the eons. Making the assumption that it has a uniform brightness is acceptable—but what if it gets brighter or fainter with age? By the 1970s, these assumptions had to be dramatically altered. Just as astronomers centuries ago were forced to change their models of the universe as soon as they realized that stars were not immutable, current-day observers have come to understand that galaxies are much more than tiny "spots in space." And this realization ushered in a new era in extragalactic research.

Richard Larson, a Yale University astrophysicist who has long thought about the origin of galaxies, once noted: "As recently as [the late sixties], the prevailing view was that all (or most) galaxies formed within a short time more than ten billion years ago and have remained essentially unchanged in structure since then. However, we are now led to a more dynamic picture in which the properties of galaxies are not immutable but can be altered, sometimes drastically, by interactions with neighboring matter." Astronomers were partly victims of their own improved instrumentation. As telescopic imaging capabilities became more sophisticated, they began to see that galaxies can exhibit diverse dispositions. While the majority of galaxies do move serenely through the cosmos, conserving their energy and changing very little over billions of years, a certain portion have been observed to behave more recklessly. It is now known that galaxies can collide, merge, sideswipe one another, or gobble up unwitting passersby. The resultant galaxywide temblors often trigger the birth of millions of stars. It is a wondrously new and invigorating picture of extragalactic affairs—one that forced Sandage's classical tests for determining the universe's fate to be temporarily shelved. "The universe has turned out to be perverse," galaxy expert John Huchra of the Harvard-Smithsonian Center for Astrophysics wryly remarks. "It's not going to give us simple answers."

Many astronomers credit Larson's former colleague, Beatrice Tinsley, with taking the first crucial step in this reappraisal. Before being felled by cancer at the age of forty in 1981, Tinsley meticulously modeled the way entire populations of stars age within a galaxy, consequently changing the galaxy's luminosity over time. Like so many before her, she was motivated primarily by the desire to use galaxies as cosmological probes. Tinsley concluded that galaxies can undergo substantial evolution internally, much more than had been previously assumed. She surmised that young galaxies probably started out very bright and blue, when resources were at their peak, and then gently faded as billions of years went by. Picking up on this lead, other astronomers began to wonder how galaxies might change, either dim or brighten, owing to influences originating from *outside* the galaxy. Is a galaxy shaped primarily by nature—the celestial "genes" it inherits at birth, such as the mass, size, and rotation of the original protogalactic cloud—or by nurture—the galaxy's later encounters with neighboring galaxies and the intergalactic medium?

This puzzle is highlighted by a significant segregation in effect throughout the cosmos. In the late 1970s, Dressler rigorously confirmed an intriguing pattern that pioneering extragalactic astronomers had noticed early in their observations: While spirals are most commonly found in loose, irregular clusters of galaxies, as well as "in the field" (the more sparsely populated or rural regions of space), gas-depleted ellipticals and ruddy S0 galaxies tend to predominate in the "urban centers," the cores of densely packed clusters, where hundreds to thousands of individual galaxies are gravitationally bound together. In the field the mix of galaxies is about 70 percent spirals, 18 percent ellipticals, and 12 percent "transition" types such as the S0 disks. In the central regions of rich clusters, though, the population breakdown is more like 20 percent spirals, 40 percent ellipticals, and 40 percent S0 types.

Were ellipticals and S0s selectively born in the more densely populated regions of space? Some of them do appear to be very, very old, as old as the universe itself. Or were they created later, when an initial population of gas-rich spiral galaxies fell into a cluster and merged? Conditions within a cluster of galaxies must exert some powerful influences on a galaxy, perhaps changing both its structure and composition.

Such questions are reminiscent of the "nature-versus-nurture" debate in human psychology circles. When it comes to galaxies, astronomers have a hunch that both nature and nurture contribute. "It's got to be a combination," says Dressler. "One could be the 'cake,' the other the 'icing.' " Conventional wisdom has held that giant elliptical galaxies, with their ancient red stars, formed during major and very rapid bursts of star formation, which exhausted their supplies of gas early on, before any matter could settle into a disk, and that spiral galaxies, coalescing out of protogalactic clouds with higher rates of spin, were somehow able to collapse in a more leisurely fashion, holding onto a cache of raw material that condenses into hot new stars to this day. But how much of a contribution does the environment make in a galaxy's ongoing evolution? This question still stumps astronomers. Is it the cake or merely the icing? Or are nature and nurture comparable in influence—the meat and the potatoes in a galactic stew?

MIT's Alar Toomre provided part of the answer by casting his mathematical eye on peculiar galaxies, so named because of their bizarre appearance. Many of these strange galaxies are part of multiple systems, where members are bumping into one another. The strongly interacting galaxies, most often pairs, reveal themselves by exhibiting structures, such as plumes, wisps, and streamers, that are seen nowhere else in the universe. "Almost every crowd includes a few charming eccentrics or confounded exceptions. This is true of the 'crowd' of galaxies," says Toomre. A few percent of the galaxy population are classified as peculiar. Toomre's early devotion to these galactic oddballs, simulating a series of galaxy–galaxy encounters on a computer in collaboration with his brother Juri, helped spawn a new astronomical industry. Such simulations offer astronomers a convenient laboratory to study the types of interactions that more ordinary galaxies may experience at least once in their lifetime, maybe even more often.

Hubble noticed a few distorted galaxies during his initial surveys of the sky but paid scant attention since it took all his efforts just to systematize ordinary galaxies. Moreover, with the instrumentation then available, the unusual structures were barely discernable. And for a long time, astronomers staunchly accepted the belief that the odds against two galaxies meeting one another in the vastness of space were huge. But by the 1950s and '60s, astronomers such as B. A. Vorontsov-Velyaminov in the Soviet Union and Halton Arp of the Hale Observatories were able to catalog up to several hundred peculiar galaxies—a number that the Toomres called "an untidy and baffling multitude." The evidence convinced Soviet astronomers (in fact, much earlier than the Western astronomical community) that a galaxy could indeed be changed by its environment.

"If galaxies were totally free in space, they would not have a snowball's chance in hell to encounter one another. But we have learned that galaxies come in groups and pairs, and this makes for a much better chance for them to collide, to form tails and bridges, and to punch holes in one another," says Toomre, whose office walls at MIT prominently display many engaging examples. The galaxies' twisted shapes clearly demonstrate the damage that gravity can inflict whenever two galaxies brush past one another. It's as if galaxies were born to interact, not unlike the way comets, traveling in highly eccentric orbits, periodically approach the Sun to blaze forth in full though brief glory.

Peculiar galaxies are often given charming names (the "Playing Mice" and the "Toadstool" are two examples) that refer to their picturesque shapes. One of the best-studied specimens is known as the "Antennae" because of the interacting galaxies' marked resemblance to insect feelers. The very faint filaments were first noted as far back as 1917. The two galaxies involved, NGC 4038 and NGC 4039, look just like the rabbit ears perched on many television sets. Yet at one time, NGC 4038 and NGC 4039 were separate and, presumably, quite ordinary galaxies. But 700 million years ago, the spiraling disks started to sweep past one another, like two majestic cruise ships passing in the night. The Toomres could see from their graphic computer simulations that colossal gravitational tides, analogous to the Moon's daily tugging at the Earth's oceans, were engendered by the encounter. As a result, two enormously long and narrow tails of material, one from each galaxy, were pulled outward as the galaxies passed by one another. Each appendage is more than half-a-million light-years long. Some observers like to joke that they've caught the galaxies "in flagrante delicto." The cataclysmic meeting not only dragged hordes of stars out of each galaxy but the resultant shock waves sparked a round of star formation that continues to this day. Obviously the original galaxies will never be quite the same again. Interactions are most disruptive when the two galaxies are moving by one another fairly slowly. Too quick a passage and the galaxies experience hardly a quiver, not having enough time to appreciably affect each other gravitationally.

But passing galaxies don't always miss one another altogether. A galaxy known as the "Cartwheel" demonstrates this other type of galactic encounter. As its name implies, the Cartwheel displays a prominent hub, with spokes of stars and gas that extend to an outer circular ring. Toomre says the odd structure also reminds him of the ripples that develop after a pebble has been dropped into water, and for good reason. It is believed this galaxy got its shape when a massive intruder poked right through its middle about a quarter of a billion years ago. The beautiful Whirlpool galaxy, the first nebula to have its spiral structure discerned by Lord Rosse, was also the victim of a glancing blow. Not too long ago, several hundred million years or so, a small irregular galaxy swung over the famous face-on spiral like a gymnast flying over a pommel horse. This flyby caused a diffuse plume of material to be flung back from the receding interloper, toward the Whirlpool's outer edge. The Whirlpool, in turn, spewed a long arm of hydrogen gas into the interloper's trail. From our vantage point, the Whirlpool resembles a concerned master attempting to keep a tight rein on his errant puppy. In time, the intruder will probably return, hauled in by the Whirlpool's mightier gravitational pull.

Low-level interactions between galaxies go on all the time, even in our own backyard. Right now, the Magellanic Clouds are leaving a stream of debris in their wake as they whisk by the much larger Milky Way. "Most interactions are probably just like this—dwarf galaxies, which are fairly common, encountering great big galaxies," points out Kitt Peak astronomer John Gallagher, who has studied such skirmishes. "Like a freight train slamming into a small car, the dwarf is probably more damaged. The freight train may not even know it has hit anything, while the small car is simply cut to shreds."

François Schweizer, who is with the Department of Terrestrial Magnetism of the Carnegie Institution of Washington, became quite enchanted with the "animals from the zoo of Alar Toomre," as he likes to describe peculiar galaxies. Considerably influenced by the computer simulations conducted by the Toomres in the early 1970s, Schweizer decided to hunt down the most dramatic interaction of them all: the actual merger of two galaxies. The Toomres had suggested that galaxy mergers were all but inevitable, but definitive proof escaped the notice of astronomers. Some doubted that mergers even occurred. "Astronomers were willing to accept that tidal interactions took place, but mergers were another thing," recalls Schweizer.

While he was a staff member at the Cerro Tololo Inter-American Observatory in Chile during the mid-1970s, Schweizer spent many lonely nights at the observatory's 4-meter telescope taking dozens of deep photographic exposures and spectrograms of several potential merger candidates. NGC 7252, an isolated and very disturbed galaxy in the southern sky, became his prime suspect. The smoothness of its central round body is marred by subtle rippling and powerfully arcing loops, and there is no neighbor nearby to accuse of trespassing. NGC 7252 came to be known as the "Atoms-for-Peace" galaxy because it vaguely resembles the old-fashioned symbol for an atom (several oval lines encircling a tightly packed atomic nucleus). Actually, the curved filaments coming out of its fuzzball-like center make it look more like a crumpled spider. Two slender and very faint tails jut out of the galaxy entirely, forming a sort of check mark on the sky.

"One of the main lessons that the Toomres taught us," says Schweizer, "is that when you have two long tails—and these tails are long; one would extend twice the distance from here to the Magellanic Clouds—then there are two participants." By carefully measuring the velocities of gas and stars within his suspect, Schweizer became convinced that NGC 7252 was born of the turbulent union of two disk galaxies. He saw vestiges of the spinning disks hidden within the chaos. The tails and loops were definitely not explosive ejecta, as some observers had alternately proposed. "It may have been a double-galaxy system that went through two or three encounters until the pair ultimately stuck like glue," says Schweizer. "The dynamical friction between the two galaxies got so strong that they simply tumbled together, and we're witnessing the result a billion years later."

Schweizer checked and double-checked his evidence for a full six years before publishing his verdict, since there was great resistance to the idea. But he didn't despair. "Thank God we're not all suggesting the same hypotheses," he says. "The beauty of science is that we all complement one another. If we all agreed and worked on the same theories, then science would not move nearly as fast. It's a battle of the minds, of ideas, and of approaches." Ironically, the concept of galactic collisions and mergers is not only routinely accepted now, it is almost required to explain how the universe generates some of its most energetic galaxies.

Schweizer expects that eons from now the messy conglomeration of stars called NGC 7252 will settle down to become an ordinary, run-of-the-mill elliptical galaxy. The Toomres, in fact, have boldly suggested that many of the ellipticals we see in the universe were formed by the merger of disk galaxies. Taken to the extreme, this model of galaxy evolution pictures only spiral galaxies forming at the dawn of creation; ellipticals would have come later as the disks started smashing into one another. It's a highly controversial proposal, and one with many problems. (How, for instance, do those merging spirals lose their spin? Ellipticals have little angular momentum.) But it's a model that cannot be immediately dismissed. Collisions were probably more frequent in the past when galaxies were crowded more closely together. Interestingly enough, Schweizer examined a sample of thirty-six fairly isolated ellipticals and discovered that half did seem to exhibit the telling symptoms—ripples, tails, and shells—of being the end products of disk galaxies merging. While most astronomers are not convinced that all ellipticals were created in this manner, Schweizer's data seem to prove that some were. This means that the environment can indeed, at least occasionally and sometimes drastically, control a galaxy's destiny.

Even the Milky Way will eventually drag the Magellanic Clouds inward and gobble them up, brightening itself considerably in the process. Future Milky Way residents won't have to worry about hordes of stars crashing into one another during the collision—stars are just too far apart—but innumerable gas clouds will get compressed and shocked as the Clouds plunge inward, triggering a veritable deluge of star formation. The most massive stars that are spawned in this burst will rush through their life cycles and explode as dazzling supernovae. The resulting flood of cosmic rays could pose quite a danger to nearby life forms. But don't worry just yet. This isn't expected to happen for several billion years.

Mergers are not restricted to disk galaxies. Existing ellipticals, too, can experience an occasional intrusion. Two very agitated and strong radio-emitting ellipticals, Fornax A and Centaurus A, show all the signs of having recently devoured some gas-rich spirals. The disks of the spirals, after being stretched and distorted, appear to have wrapped themselves around the central cores of the ellipticals, like stripes coiling around a candy cane. A trail of delicate ripples and enormous loops of gas serves as the telltale footprints of the fatal collision.

The galaxies with the largest appetities, though, live in the hearts of dense, rich clusters of galaxies. These gargantuan ellipticals, labeled cD in astronomical catalogs (for supergiant D-type galaxy), are puffed up to sizes that could accommodate a dozen or more Milky Ways very comfortably. In the mid-1970s, Ostriker and Scott Tremaine suggested that the cDs acquired their ample figures by gorging on lesser neighbors that happened to stray too close. These poor unfortunates, caught in a gravitational web, supposedly spiraled inward and got stripped of their stellar layers. Each cD, as a result, became very bloated and increased its luminosity appreciably. Over a few billion years, a typical cD could have swallowed several dozens of large and small galaxies. They certainly stand out in the universe, being both the brightest and largest members of their clusters. Ostriker made the idea of mergers much more palatable with this proposal. The strongest evidence in support of such galactic cannibalism comes from inspecting the innards of these supergiant ellipticals. Some observers believe they have seen the remains of the victimized galaxies settling in cD "stomachs" long after the feast. Some cD galaxies display several remnant nuclei. "A case of the rich getting richer at the expense of the poor," says Ostriker. Galactic cannibalism is more than a merger; it's a cosmic takeover.

As additional refreshment, cD galaxies have been slurping up a bit of the gas that permeates rich clusters of galaxies. Soon after x-ray space telescopes were launched into orbit during the 1970s and focused their eyes on extragalactic space, astronomers became very aware that many clusters of galaxies are immersed in a tenuous sea of hot, x-ray-emitting gas. This gas may help researchers track down the origin of galaxies.

Mapping how the density of the gas varies—where it is highly concentrated; where it is not—is enabling astronomers to trace the underlying structure of a cluster, a framework that is invisible to optical telescopes. "The density of this gas is very low," points out x-ray astronomer Christine Jones of the Harvard-Smithsonian Center for Astrophysics. "There are only about thirty particles in each cubic foot of space, and this density smoothly falls off as you proceed to the outer borders of the cluster." Yet, considering the vast volumes of space taken up by a cluster of galaxies, there is just as much mass tied up in the gas as there is in galaxies, which means its effects are wide-ranging. High-energy astrophysicist Wallace Tucker has said that it can play havoc with a galaxy. "So much havoc," he has pointed out, "that if our galaxy had happened to be in the central regions of a cluster rather than where it is, the Sun probably would never have formed."

A large fraction of this hot cluster gas seems to have come from the galaxies themselves, blown out through the successive blasts of high-powered supernovae. But this pool of superheated material is supplemented. A cluster of galaxies is so crowded that encounters between galaxies are likely to rip away some more material, which is added to the mixture. And if a gas-rich spiral should fall in toward the center of a tightly bound cluster, some of its gas and dust could be either stripped away by ram pressure as the galaxy courses through the relatively dense intracluster medium or "evaporated" out as the relatively cool galaxy comes in contact with the hundred-million-degree gas found in the cluster's inner regions. At the Center for Astrophysics, Jones and her collaborator-husband William Forman have examined many x-ray portraits taken by the *Einstein* observatory and in several clusters have seen plumes of gas strewn behind fast-moving galaxies. It's as if the gaseous tendrils were strands of hair being swept back by a celestial gale. Spiral galaxies, robbed of some of their precious raw materials in this manner, are left to a dreary fate: evolution into smooth disks devoid of spiral arms. This could be one way that some of the spiral-less S0 galaxies, huddling in the cores of rich clusters, came into being. With their gas supplies depleted, bright new stars would no longer be born in the spiral arms, which are essentially the luminous crest of a compression wave moving through the disk, and thus the arms disappear. With the cutoff in the production of young blue stars, such galactic disks take on a more reddish and aged cast. This is why our Sun, a second- or third-generation star, might not have had the opportunity to form if the Milky Way had found itself plunging through a rich cluster. The Milky Way's gaseous resources would have been blown outward into intergalactic space long before the Sun and its family of planets had a chance to coalesce.

Or maybe a spiral galaxy plunging through a cluster prematurely exhausts its gas by an overwhelming episode of "starbursting." "There is reason to believe," notes Dressler, "that the interstellar medium within a spiral gets compressed as it plows through a cluster. As a result, star formation would proceed at a much higher rate, until it ultimately depletes all of the spiral's gas. It's like running a film very fast: The spiral is hastening its natural evolution. Similarly, a pair of galaxies constantly encountering one another can accelerate the star-formation process. It has taught us that some galaxies have not been dormant—coasting along at low ebb over the last ten billion years."

As the name implies, a starburster is a galaxy gone frantic, at least temporarily. In many cases, a merger or collision serves as the fateful trigger.* A starburster may even be fueled by material stolen away from its passing companion. Starbursting, in fact, can be one of the more visible consequences of a galaxy–galaxy encounter. Jostled and shocked by the gravitational tugs of a sideswiping neighbor, the starburster's interstellar clouds are compressed more rapidly than usual, leading to the production of millions of new stars. A large part of the superaccelerated production occurs in the galaxy's center. M82, a prototypical starburster whose wildly disturbed profile was once a longstanding astronomical enigma, was explosively awakened several tens of millions of years ago by the close passage of the giant spiral M81. All sorts of stars, especially very massive ones, are manufactured at a frenetic pace (astronomically speaking). Stellar explosions pop off like Fourth of July fireworks. Instead of a supernova exploding every thirty to fifty years, as in the Milky Way, the most violent starbursters can produce detonations yearly. The galaxy is stirred, shocked, and shaken, until the instigator, the intruding companion, steals safely away. Astronomers believe

*Isolated starbursters have been detected. "It's easy to see how collisions would trigger an episode of starbursting," says Arizona astronomer George Rieke, "but we are not sure at all how the isolated cases are started."

that any one episode of starbursting probably lasts no longer than a hundred million years. Some observers estimate that about a tenth of the galaxies in the universe either have experienced or will experience a particularly intense starburst at least once in their lifetime, though lower levels of excitation do occur in a larger fraction. The mysterious activity in the central regions of our own Milky Way is one such low-level case.

The fact that galaxies can occasionally experience very short but intense bursts of star formation was established most effectively by the IRAS satellite. When the sensitive infrared instrument scanned the heavens, it detected a number of galaxies that emit up to a hundred times more *infrared* energy than the Milky Way does at *all* wavelengths. One of the most remarkable is a galaxy known as Arp 220 (from Arp's *Atlas of Peculiar Galaxies*), which seems to be either merging or colliding. Aside from a fuzzy distortion, this galaxy was fairly innocuous through the lens of an optical telescope; no one took much notice. But in IRAS's eyes, Arp 220 shone with a blinding infrared light. It is believed that the galaxy's radiations are being absorbed by numerous dust clouds churned up during a powerful burst of star formation (400 times the Milky Way rate of one or two solar masses a year) and then reradiated as enormous amounts of heat energy. Arp 220 and others like it could be mimicking the way galaxies looked more than ten billion years ago when stars first turned on. A starburster could thus serve as a model for one of the most elusive animals in the celestial wilderness—the "primeval galaxy."

Astronomers have been able to glean a tremendous number of clues concerning the origin of galaxies from studying our closest neighbors. Nearby galaxies offer observers the opportunity to compare the salient features of an elliptical with that of a spiral in minute detail and to perform laborious autopsies on the scattered remains of interacting galaxies. A standard joke among particle physicists is that their attempts to understand the construction of elementary particles by slamming the particles into one another at near-light speeds is much like trying to understand the workings of a watch after smashing it with a hammer. Inspecting the wreckage of a galactic collision offers a similar pathway toward learning how galaxies might have evolved over time. But certain questions will undoubtedly remain unanswered until astronomers can push back the limits of their vision to that remote epoch when galaxies first coalesced out of the primeval "soup." As theorist Peebles once noted, "If we saw what young galaxies [were] like it would settle a lot of arguments."

"If we looked back twelve billion or more years and found that everything was a disk," adds Dressler, "then we'd have pretty strong evidence that it was later interactions [nurture] that caused things, like ellipticals and S0s, to come into existence. But if we found that there were big bulges already forming at the centers of regions destined to become clusters, then we'd know that initial conditions [nature] were the deciding factors."

The solution seems simple enough: Just turn the cosmic clock back some ten to fifteen billion years and take a look around. "Astronomers have an advantage over other historians," points out Dressler. "We can observe history *directly.*" Since light travels at some 186,000 miles per second, the farther away an object is, the longer its light must take to reach our telescopes. Therefore, as pointed out earlier, when astronomers peer out to the boundaries of the visible cosmos, they travel back in time. But this journey is an extremely difficult one. The deeper astronomers peer into extragalactic space, the more they must strain their telescopic "eyes." Galaxies appear dimmer, smaller, and fuzzier. At a certain point astronomers become downright myopic; they are unable to distinguish whether a galaxy is a spiral or elliptical at all, although educated guesses can be made by looking at the galaxy's spectral features. Advanced electronic detectors are making it possible to obtain, fairly regularly, spectra of galaxies out to eight billion light-years—halfway back to the beginning of the universe.

Yet the origin of galaxies continues to elude astronomers. What Hubble wrote in the 1930s holds true for today: "With increasing distance, our knowledge fades, and fades rapidly. Eventually, we reach the dim boundary—the utmost limits of our telescopes. There, we measure shadows, and we search among ghostly errors of measurement for landmarks that are scarcely more substantial."

Only a few handfuls of astronomers have chosen to make the farthest reaches of the cosmos their life's work. It takes a special breed of astronomer. Years of survey work are often required before conclusions in this field can be reached. Answers are not guaranteed, and resources are limited. Collecting data from the edge of the visible universe requires the use of the world's largest telescopes, whose number are few, and the observatories receive many more requests for time on the instruments than is available.

David Koo and Richard Kron are two such astronomers devoted to exploring the remotest nooks and crannies of the universe. Their primary instrument is the 160-inch Mayall telescope at the Kitt Peak National Observatory. During their graduate studies in the 1970s, the two eagerly watched the telescope and its stately white dome being assembled on the southern Arizona mountaintop, for they knew it was going to be their time machine into the distant past. Kron is now based at the University of Chicago's Yerkes Observatory; Koo is with the Space Telescope Science Institute in Baltimore. But the close collaboration that began in their student days continues. They expect to spend years, even decades, surveying several small regions of the sky and wringing every bit of data they can out of the photographic images and spectroscopy of those sectors. "We're fond of faint, fuzzy dribbles of data from afar," says Koo. "We don't want to deal with the well-known." Like oil-well drillers, they will gradually probe deeper and deeper into space, further and further back in time. Eventually, they hope to be able to make general comparisons of the galaxies of yesteryear and the galaxies of today. They don't anticipate immediate answers to their questions; they know that patterns will emerge and connections will be made only as more and more evidence is painstakingly gathered.

So far, Koo and Kron have gotten an impression that distant field galaxies were somewhat "bluer" in the past. Is this just the collective result of stars being younger and more vigorous in each distant galaxy? Or does this mean there were great upheavals taking place in that far-off time: galaxies colliding, merging, and starbursting, thereby changing their morphology more appreciably than today? Were there simply more blue, gas-rich spirals around? "Or is the 'blueness' of a faraway galaxy an illusion," asks Koo, "the result of our ignorance concerning the general properties of galaxies?" No one knows for sure yet. "We have to give the universe credit for being subtle," adds Kron.

At least a few distant (and hence ancient) clusters of galaxies appear to share this stronger tinge of blue. In 1978, when Harvey Butcher and Augustus Oemler analyzed the light emanating from two clusters situated more than five billion light-years away (hence five billion years back in time), they discovered that the far clusters radiated more blue light than the dense and somewhat reddish clusters near us today. This has been dubbed the "Butcher-Oemler effect." The two astronomers surmised that blue, star-forming disk galaxies were more common at that time and that some kind of intracluster activity must have exhausted these galaxies' gas as the clusters evolved to the present day. Dressler and James Gunn of Princeton obtained spectra for several dozen galaxies in other far clusters and saw evidence that the cluster galaxies were more agitated; there was heightened star formation and increased activity in the nuclei of the galaxies, far more than seen today. All of this hints at dramatic developments occurring in galaxy evolution over the last 5 billion years.

Some astronomers are tempted to conclude that this noticeable decline in activity over time proves that galaxies in general are rapidly "running down," as less and less gas becomes available for their star-forming needs: a growing cosmic ennui. But such a conclusion cannot be rushed. Distant galaxies are slow to unveil their secrets. The evidence, in fact, has been contradictory. Koo and Kron detected a faraway cluster that was very red in its day, implying that little was changing five billion years ago. Any prodigious bursts of star formation in that cluster must have occurred much earlier and very abruptly, for the galaxies in it had already settled down to a fairly quiescent life-style. "This cluster may be a maverick," cautions Koo. But, then again, it may not be. Surveys by other groups only add to the confusion: Some perceive great upheavals a third of the way back to the Big Bang; others swear they see little activity at that time. "It's like that old story about several blind men feeling selected parts of the elephant," muses Koo. "One feels the tail and concludes the animal is a snake. Another feels the leg and says it is definitely a tree. Their stories appear to conflict, but they just haven't seen the whole elephant yet. The universe is obviously displaying a whole range of behavior."

Astronomers strongly suspect, though, that they are on the threshold of discovery. It's as if they were historians who had been constrained to limit their studies only as far back as World War II, but are now being allowed to examine records that go all the way back to the days of the Roman Empire and beyond. The Hubble Space Telescope, three hundred miles up and free of the Earth's shimmering atmosphere, will be able to resolve spirals and ellipticals billions of light-years more distant than current instruments are capable of doing; meanwhile, new, advanced telescopes on the ground will determine the galaxies' redshifts more quickly and easily than is presently possible. Combined, the results could be as revolutionary as the discoveries made possible by the 100-inch telescope on Mount Wilson at the start of this century—and that instrument redefined the universe.

Hyron Spinrad, a veteran at gazing at the universe's outermost boundaries, compares these endeavors to those of a pole vaulter. "Every year," he says, "we raise the bar a little higher." In the early days of extragalactic astronomy, the bar stood at around 150 million light-years. Between 1929 and 1954, owing largely to the efforts of Milton Humason, the bar quickly moved out to 2 billion light-years. Today, galactic redshifts are regularly detected out to 8 billion light-years, and once in a while even beyond that. Whether such efforts succeed, it's been said, is partly a function of technological innovation and partly a matter of sheer persistence. Spinrad and his colleagues at Berkeley, for instance, have been known to look at a particularly faint galaxy over twenty nights, adding together each night's meager data to obtain just one fuzzy smear or a few spectral features on their image detector. The Palomar sky survey, an extensive photographic atlas of the heavens completed in the 1950s, doesn't even record the distant galaxies that the Berkeley team is after, since these dim objects are millions of times fainter than the stars we see with the naked eye. To know where to look, the Berkeley astronomers must depend on radio astronomy. The distant ellipticals that they study, too faint to be seen directly through the telescope, emit intense radio waves from their centers. "The radio signal being sent out by the galaxy tells us where to point the telescope, or else we'd never find them," explains Spinrad. "The stakes are getting higher as we go out to the really faint stuff. About a third of our sources turn out to be washouts. Sometimes they're close by; sometimes we never find out what they are. But then we try again, sinking our teeth into these things like bulldogs."

A bit of luck doesn't hurt, either. While trying out a new filter on a familiar quasar just for fun (a spectral filter designed to study Halley's comet as it flew by the Earth in late 1985 and early 1986), Stanislav

Djorgovski and Spinrad were surprised to find that the quasar had an unexpected companion. It wasn't much of a companion, no more than a soft-looking smudge off to one side of the quasar, but its redshift told them the galaxy was located some 11 billion light-years from Earth. The photons that left this galaxy to place their mark upon terrestrial detectors started on their journey at a time when the Milky Way was forming. As of this writing, it is the farthest relatively normal galaxy yet detected. (Quasars have been found farther out, but they appear starlike.) Eventually, the Berkeley group hopes to reach all the way back to the era, still hidden, when primeval galaxies experienced the first pangs of stellar birth, the time when an archipelago of island universes emerged from the chaos of the newborn universe.

Spinrad concedes there's a pitfall in these efforts at the far frontier. He and the others in this line of work could be observing only the most ostentatious galaxies, the ones with "overactive thyroids." As astronomers go farther back, the brightest galaxies (and thus perhaps the most unusual) tend to stand out in the crowd. Some of Spinrad's most distant galaxies do look elongated, as if they are colliding with gas-rich neighbors. Such specimens could be a lopsided sample of what's really going on at the universe's periphery.

So the Spinrads, Dresslers, Koos, and Krons must keep returning to their telescopes if they are to piece together the most complete evolutionary history of galaxies possible. Humans go through distinct stages: infancy, adolescence, middle age, and old age. Does a galaxy? The one lesson that astronomers have assuredly learned is that a galaxy is a complex creature. It can brighten and dim; merge with other galaxies; or remain a wallflower lifelong. Although this relatively new view of galaxies was a surprising one to astronomers, it wasn't as perplexing as another picture of galactic behavior that was also beginning to emerge.

To this point in the story, we have discussed only the more "normal" activities of galaxies. The galactic events described so far are what one might expect for a collection of billions of individual stars under the control of gravity. Whether the galaxies are colliding, merging, or maintaining majestic aloofness, the energies that they release can be explained as some variation of a well-known stellar process. Even the monstrous energies given off by certain starbursters require nothing more exotic than a resounding chorus of supernovae spiritedly exploding. But at one point astronomers were forced to confront the fact that a single galaxy's nucleus can, during special moments in its lifetime, either scream out in a loud radio voice or burst forth with the combined brilliance of hundreds of galaxies. The evidence was startling and upsetting. Try as they might, astronomers could not balance the ledgers. The tremendous energies spewing from these newfound objects could not be accounted for in terms of ordinary stars burning or dying. It has been said that to explain their bizarre properties "is one of the greatest challenges that has been presented to science in modern times." These cosmic beacons are the enigmatic radio galaxies, active galaxies, and quasars—a cosmos turned rogue.

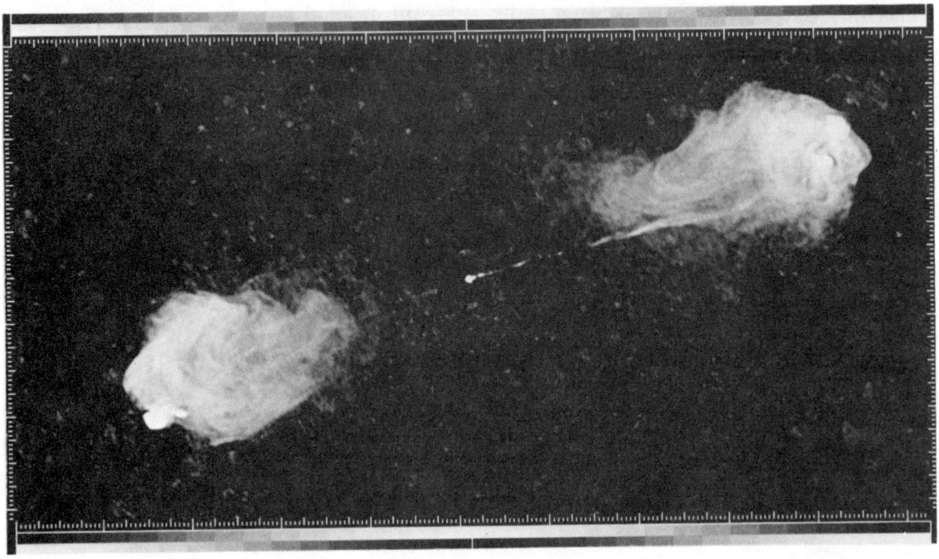

Nature's welding torch. Two thin, straight beams of radio-emitting plasma spew from a mysterious "engine" situated in the center of the giant elliptical galaxy Cygnus A (which is not visible in this radio "picture"). The cosmic jets bore through intergalactic space until they disperse, creating a pair of gigantic radio lobes. *Courtesy of the National Radio Astronomy Observatory, operated by Associated Universities, Inc., under contract with the National Science Foundation.*

6

GALACTIC DYNAMOS

There's a monster lurking in the heart of these galaxies, and food is being fed to the monster.

—Richard Green

At the eastern edge of San Jose, the hub of California's computer and electronic industries, Donald Osterbrock turned his car onto Highway 130 to start a long climb up the slopes of Mount Hamilton,* where gnarled oak and pine trees fill the mountain's furrowed sides like well-sprinkled seasoning. He was starting an 18-mile journey, complete with 366 hairpin turns, familiar to four generations of astronomers. The original roadway, once a simple dirt lane wide enough for horse-drawn wagons, was built by county laborers in 1876 because James Lick, a

*Named after an adventurous San Jose minister who accompanied two surveyors on a climb up the mountain in the summer of 1861. Upon reaching the summit before his companions, the Reverend Laurentine Hamilton shouted, "First on top!"

wealthy San Francisco real estate investor, wanted to construct "a powerful telescope, superior to and more powerful than any telescope yet made," on Mount Hamilton's 4,200-foot-high summit, the tallest peak in the Diablo Range that borders the Santa Clara (alias Silicon) Valley. The Lick 36-inch refracting telescope, completed in 1888, held its title of most powerful telescope for nearly a decade and is still the world's second largest refractor, an instrument that focuses light, not with mirrors, but through a pair of yard-wide lenses like some kind of colossal spyglass.

But Osterbrock was not headed for the 36-inch refractor that cool autumnal day. While the Lick telescope can, even after a century of use, provide important data on some of the nearest stars and planets, Osterbrock had his eye on objects beyond the reach of the refractor—a somewhat rare species of spiral galaxy, first spotted in the 1940s, which violently emits the energy of dozens of galaxies from a bright, compact core. For this job, Osterbrock needed Lick Observatory's largest telescope, a reflector whose 120-inch-wide mirror sits at the lower end of a 50-foot-long cylindrical framework of steel tubing, painted a vibrant yellow, that moves on a massive mount that resembles a giant two-pronged fork.

At eight o'clock that evening, while a brisk 20-mile-an-hour wind blew across Mount Hamilton's summit, the telescope operator typed a series of commands at his computer terminal, directing the telescope to its first target. He, Osterbrock, and two student assistants were comfortably ensconced, off to one side of the telescope, in a room reminiscent of a NASA mission control center with its glimmering array of gauges, monitors, and electronic instrumentation. Contrary to romanticized notions of astronomical observing, of gazing at the heavens from a dark, cathedral-like dome, astronomers no longer peer through a telescope's eyepiece to keep a galaxy or nebula in sight. Osterbrock, a senior astronomer who got his start by helping William Morgan make the first maps of the Milky Way's spiral arms, learned his craft on a telescope that is now on display in a museum, and he is hardly nostalgic over that bygone era: "During the wintertime, bundled in layers of clothing, I'd freeze sitting at the prime focus for hours and hours. And then I would have to dash off to the darkroom in the unlighted dome and develop the photographic plates in order to see what the data looked like. . . . I even bumped my head on the telescope."

Nowadays, photons entering the giant reflector from far-off galaxies are directed to a television camera, which displays the celestial image on a TV screen prominently mounted in the control room. Osterbrock's first target, one of a dozen he would study during that nightlong vigil, was a very special member in the cosmic menagerie. In catalogs, it is listed as NGC 6814, a swirling spiral galaxy whose light started on its journey to Earth at the time the dinosaurs died. Osterbrock didn't see the spiral's disk, only a grainy spot of light that flickered and danced ever so slightly on the monitor. And like a veteran video-game player, he gently tweaked a joystick to position the brilliant dot so that the galaxy's light, after entering the telescope, would be sent through a spectrograph and broken down into composite wavelengths, much the way a prism will create a palette of color out of sunlight. These spectral features are to astronomers what cardiograms are to heart specialists: a unique pattern that reveals information not apparent to the eye. Since every chemical substance absorbs and emits radiation at specific wavelengths, and thus has its own distinct "fingerprint," spectral analysis enables astronomers to deduce a celestial object's chemical composition, temperature, and internal motions.

Sensitive detectors recorded the spectrum of NGC 6814 for fifteen minutes. Shortly afterward, the pattern was displayed on a computer monitor. The spectral graph looked like a ragged mountain range, indicating which wavelengths or colors of light the galaxy was emitting (the hills on the graph known as emission lines) and which it was absorbing (the valleys, or absorption lines). The positions of the peaks told Osterbrock's team that such elements as nitrogen, oxygen, sulfur, and iron were being heated within the nucleus of NGC 6814 to tremendously high temperatures, causing the gases to fly outward at hundreds, even thousands of miles per second. "Normal galaxies just don't do this," says Osterbrock, a former director of Lick Observatory. "Some kind of engine has to be residing in this galaxy sending out a lot of energy that excites the gas clouds we're detecting."

American astronomer Carl Seyfert first identified this strange behavior in 1943 during a routine survey of spiral galaxies. Here were objects, Seyfert noted, that looked very much like the Milky Way, except for one crucial difference: While their disks appeared perfectly normal, their bulbous cores were outshining our home galaxy a hundred times over. Despite the tremendous technological advances made since Seyfert's day, these queer spirals retain much of their aura of mystery; their enigmatic "engines" have yet to be seen directly. The detection of Seyfert galaxies was the very first hint to astronomers that galaxies can be more than mundane collections of stars slowly burning in a tranquil cosmos. Galaxies, it was being revealed, have an explosive edge to their personality. The energies being released are so enormous that astronomers suspect that this fierce activity has to be an aberration rather than a lifelong condition. Quasars, caught at the edge of the visible universe, are the most violent expression of this trait. Such galactic outbursts could be but a brief, raging temper tantrum unleashed during a relatively short moment in cosmic history. As we saw earlier, the Milky Way may have flared up in a similar manner, or may without warning in the future. Why? That is one question that draws astronomers like Osterbrock back to their telescopes again and again. Seyfert's discovery was the first stroke in an ongoing astronomical revelation: the realization that violence, not serenity, reigns supreme in the cosmos, that what is out there can be more like Mr. Hyde than Dr. Jekyll. This emerging portrait of a universe peppered with exploding galaxies was not fully unveiled, however, until astronomers in the 1950s took advantage of a brand new tool at their disposal: radio telescopes.

"I have been receiving a very weak and very steady static lately," a young Bell Telephone Laboratories researcher wrote to his father in the early months of 1932. "The peculiar thing about this static is that . . . [it] always comes from a direction that is the same. . . . Sounds interesting, doesn't it." It was more than interesting; Karl Jansky's discovery was revolutionary. The fact that radio waves were emanating from the center of the Milky Way made front-page headlines in *The New York Times* the following year, but astronomers were slow to appreciate the finding because their theoretical calculations indicated that radio emissions within the galaxy would be quite feeble.

Fortunately, Grote Reber was blithely unconcerned with astronomy's conventional wisdom. Upon hearing of Jansky's work, this Illinois radio engineer and avid ham-radio operator erected a massive steel saucer, 31 feet wide, in his backyard to detect Jansky's cosmic static for himself.* Constructed on his own time and with his own money, Reber's dishlike

*Reber's neighbors at first thought that his three-story structure was going to be used to either collect rain water or control the weather.

antenna, a design that has become radio astronomy's trademark, was the first device specifically designed to receive extraterrestrial radio waves. From his neighborhood outpost in Wheaton, Illinois, Reber produced the first comprehensive maps of the "radio" sky. He found that there were bright, blobby regions of radio emission throughout the celestial sphere, not just at the galactic center where Jansky first detected his cosmic noise. The radiations were particularly strong near the constellations of Cygnus and Cassiopeia. Not surprisingly, the astronomical community in the thirties and forties didn't rush to embrace Reber's novel view of the universe. The typical stellar observer in those days, more at home with lenses and mirrors, was wary of Reber's seemingly bizarre equipment and could not fully understand the astronomical implications. Only the intervention of a farsighted editor allowed Reber to publish his results, at first rejected, in *The Astrophysical Journal.* Since optical astronomers were still carrying on the work initiated by Galileo's scrutiny of the heavens, it was difficult for all but a few visionaries to understand that the cosmos could display a very different picture of itself in waves of electromagnetic energy other than visible light. It took World War II to cure this shortsightedness.

Dozens of young physicists and engineers in both Europe and the United States were introduced to the esoteric art of radio science while working on the development of radar during the war, and many were anxious to apply their newfound skills, once the conflict ended, to deciphering those radio signals detected by Reber. They were particularly eager to pinpoint the celestial objects or events sending out these low-frequency waves. To this vanguard, the radio sky was a blank page just waiting to be filled in. Unfortunately, the radio dishes that began to scan the sky on a regular basis after World War II had a very blurry, out-of-focus view of the heavens. There were literally hundreds of visible objects in the neighborhood of each fuzzy radio region—and no way to know which optical image matched up with the radio signal. This is because a single radio dish cannot resolve fine details and is thus unable to zero in on the exact position of a source.

How much detail a telescope can make out basically depends on how large its collecting surface is compared to the waves of electromagnetic energy being gathered. The lenses and mirrors in an optical telescope are gargantuan compared to a visible light wave, which is only a few hundred-thousandths of an inch long. This means optical astronomers, ultimately limited by atmospheric turbulence, can discern galaxies and nebulae that cover a mere 3/10,000 of a degree in the sky, less than one-thousandth the width of the Moon's disk as seen from Earth. That's enough resolution to read the label on a tennis ball bouncing in a court several miles away. But radio waves are hundreds of thousands of times longer than visible light's electromagnetic undulations, which means a radio dish has to be many miles wide to match optical resolutions. Faced with this formidable restriction, radio astronomy's usefulness seemed limited—until British and Australian researchers turned to an old yet clever means of sidestepping that resolution barrier.

They began connecting pairs of radio telescopes that eventually stood up to a few miles apart. Via cables or radio links, the waves received by each antenna were sent to a central processor and combined. This is known as interferometry because it analyzes how the waves from each antenna "interfere" once they are added together. If the radio waves gathered by each dish are "in step," they add up to a bright, strong signal. But if they're not coordinated, they destructively interfere with and cancel out each other. As the Earth rotates and the antennas wheel slowly to stay on their target, a pattern of peaks and troughs emerges. Astronomers, or their computers, can translate these squiggles into a rough image of the source—a picture almost the same as the one you would obtain if you had a single antenna many miles wide.

In the early 1950s, interferometers enabled astronomers to start pinning down the celestial objects sending out those indecipherable radio squeals. Cygnus A, one of the "brightest" objects in the radio sky, turned out to be a strange-looking galaxy located some one billion light-years away. It was pouring forth a veritable fountain of radio energy, emissions millions of times more powerful than the Milky Way's radio output. Radio astronomers would later learn that thousands of other galaxies—interestingly, most of them are ellipticals—emit copious amounts of radio energy in a similar way. Astronomers came to realize that they had identified a new species: *radio* galaxies. About 5 percent of all bright elliptical galaxies have been found to have such strong radio voices.

In regular optical photographs, radio galaxies can appear quite ordinary, even boring. In visible light, they often hide their radical tendencies quite well. But with interferometers radio galaxies reveal a complex and bewildering architecture. When they combine the optical image with maps of the radio signals, astronomers see that the visible part of each of these galaxies is but a smudge caught between two sizable lobes of radio emission. Looking like a pair of giant water wings, these lobes stretch out for hundreds of thousands of light-years beyond the visible galaxy's edge. Each lobe is really a diffuse cloud chock-full of electrons and protons moving about in magnetic fields. At first, astronomers assumed that a rapid chain of stellar explosions must have shot these two radio-emitting clouds out to either side of the galaxy, but that didn't explain how such energetic lobes could hang around for tens of millions of years. It became more and more apparent that these particular objects were emitting energy by processes not directly connected with stars, since it would take up to 100 billion suns, the combined luminosity of an entire galaxy, to energize the lobes via simple thermonuclear fusion, the standard means by which stars release energy. Instead, these galaxies seemed to be active because (a) their nuclei contained some kind of "monster," (b) things were being fed to the monster, and (c) this celestial beast somehow transferred the titanic energies it was producing far out into intergalactic space. The energetic Seyfert galaxies, it turned out, were not alone in their unusually high levels of activity.

The desire to expose the true nature of the monster lurking within these active galaxies spurred countries such as Great Britain, Australia, the Netherlands, and later the United States to build bigger interferometers, with more dishes added to the network to obtain better images and longer spacings between the antennas to resolve finer and finer details. These efforts culminated in the construction of the Very Large Array (VLA), whose twenty-seven antennas on the plains of New Mexico mimic a single radio telescope as large as the District of Columbia. This simulated dish helped astronomers confirm what some theoretical astrophysicists, among them Great Britain's Martin Rees and Caltech's Roger Blandford, had already suspected: that a sort of umbilical cord was running from the nucleus of a radio galaxy to its lobes, continually supplying the puffy blobs with energy. Indeed, images made with the VLA clearly show thin beams of energetic, charged particles shooting out of radio-galaxy cores at speeds as great as tens of thousands of miles per second.

Like extragalactic psychologists, VLA astronomers have spent many years analyzing the varied personalities of these "cosmic jets," as the beams have been dubbed. Some jets, like the two emanating from the center of 3C 449 (object number 449 in the third Cambridge catalog of radio sources), an elliptical galaxy 200 million light-years distant, can be relatively placid. "Such jets are usually seen in the less active radio galaxies," says Richard Perley, who has trained the VLA's antennas on 3C 449. "They have a lower velocity, and sort of diffuse away—like smoke rising out of a chimney, becoming unstable, and breaking up." If the galaxy happens to be moving through a dense patch of intergalactic space, the two jets can even be swept back like the wake of a boat.

But Perley's particular favorites are the jets with "punch power," as he likes to put it. These energetic beams of plasma spew from galactic centers thinner, straighter, faster, and farther than the more languid types. "Cygnus A is a classic example, a source we've resolved in marvelous detail," he says. What Perley and his colleagues see is one of nature's most wondrous shows. Like a welding torch, Cygnus A's jets bore through the thin gases found in intergalactic space "until they hit a more dense, undisturbed region of gas, somewhat like the stream from a fire hose coming up against a brick wall," says Perley. In the VLA's computer-generated pictures, this area appears at the end of the jet as a glowing hotspot of intense radio emission. "From this point," continues Perley, "the particles fly off, filling up the lobe region." In the lobes, the high-speed electrons find themselves spiraling around magnetic field lines, like cars continually rounding a bend. But where cars squeal in a turn, electrons broadcast radio waves. These were the signals that Reber had detected.

Radio galaxies were not the only exotic objects discovered with radio interferometers. By the early 1960s, as interferometry improved, large amounts of radio energy appeared to be coming from some stars, whose images in visible light were rather prosaic. Anyone coming across them in the sky would have written them off as faint, uninteresting blue stars in our Milky Way. But the intense radio signals coming out of those dim blue dots forced observers to take a second look. Upon examining the spectra of these "radio" stars, optical astronomers discovered that they exhibited spectral features unlike any star ever observed. To stellar experts, this was like riding down a familiar turnpike and unexpectedly finding all the road signs written in gibberish. The spectral features didn't seem to belong to any known chemical element.

The identity of these strong radio beacons was finally unmasked on February 5, 1963, as Caltech astronomer Maarten Schmidt sat at his desk attempting to write an article on a radio star known as 3C 273 for the British journal *Nature.* As a teenager in occupied Holland during World War II, Schmidt constructed his first crude telescope out of a discarded toilet-paper roll. The two pieces of glass held within the small cardboard tube could barely resolve a double-star system. But by the 1960s, he had graduated to the great 200-inch Hale telescope on Palomar mountain and was using it to analyze the optical counterparts to those intriguing pointlike radio sources. With the optical spectrum of 3C 273 spread before him, along with a crucial near-infrared line obtained by Caltech astronomer J. Beverley Oke, Schmidt came to recognize a familiar pattern of lines that had eluded him for weeks. The pattern resembled the sequence of emission lines associated with hydrogen, but they were in the wrong place: The spectral lines were shifted toward the red end of the spectrum by 16 percent. "You must understand," explains Schmidt. "Astronomers had never seen a *starlike* object in the sky with a large redshift. It wasn't even imagined."

Schmidt went home that night and exclaimed to his wife, "Something awful has happened at the office today"—not in the sense that his finding was displeasing, but that it was awe-inspiring. Schmidt had immediately grasped the implications of the redshift. 3C 273 was not an unusual star situated in the Milky Way, but rather a bizarre extragalactic object located two billion light-years away, rushing into space at some 30,000 miles per second as it was carried outward with the expansion of the universe. He knew that only an incredibly bright object could be visible from such a distance. 3C 273 had to be radiating the power of trillions of stars. Early on, Schmidt suspected that it was the brilliant and very disturbed nucleus of a galaxy. In one stroke, the Caltech astronomer had turned what was once spectral gibberish into a language understandable to all astronomers and increased the volume of the known universe, the region detectable by astronomy's instruments, several times over.

Earlier that day, tremendously excited by the discovery, Schmidt had conferred with his colleague Jesse Greenstein, and the two quickly confirmed that the spectrum of another puzzling radio star, 3C 48, was redshifted by 37 percent. That translates into a distance of 3.6 billion light-years. Greenstein had actually performed the calculation a year earlier but had rejected the conclusion, to his later chagrin, as too preposterous!

To distinguish them from stars, the blue, extragalactic specks were soon christened quasi-stellar radio sources (QSRS), or quasars. In some ways, it's a misleading acronym. Radio telescopes did uncover this unusual phenomenon, but Allan Sandage discovered that most quasars, about 90 percent, send out a virtual peep at radio wavelengths; these "radio-quiet" quasars give themselves away by the extraordinary ultraviolet radiation they emit. More than one astronomer has noted that they look like Seyfert galaxies with the volume turned up.

3C 273 is situated relatively close to us, as quasars go. But, in the years since Schmidt's discovery, astronomers have identified quasars out to ten billion light-years and beyond, speeding away at near-light velocities. "The spectral lines of these far-off quasars are shifted so much that they go off the scale," says Schmidt. "That's why they were a puzzle for so long." Astronomers found themselves observing cosmic processes near the edge of the universe, which revealed that the cosmos was a very different place during its infancy.

"Once we accepted the redshifts," says Schmidt, "we had to deal with entirely new questions, such as the energies." The fact that earthbound observers were able to see quasars clear across the universe meant that these objects were the most powerful denizens of the heavens. "The insult was not that they radiate so much energy," continues Schmidt, "but that this energy was coming from a region probably no more than a light-week across." Schmidt knew this because many quasars are seen to flicker; they periodically dim and brighten over a matter of weeks or days. Space x-ray telescopes have observed flickerings in quasars on a scale of hours and minutes. Since it would take considerable time for brightness to vary over a large area, astronomers have surmised that a quasar's power source is relatively small, perhaps less than the diameter of a solar system. If the engine were much larger, variations in its power output would smooth out and hardly be noticed. Yet from such a tiny region spews the energy of up to a trillion suns—the equivalent power of dozens of galaxies' worth of stars. Tapping into this cosmic dynamo for just one second would power the world for a billion billion years.

It was easier for astronomers to accept this bewildering situation once Oke and Sandage discerned similar flickerings in the nucleus of a relatively nearby radio galaxy called 3C 371. For many, this forged a vital link between quasars and more ordinary galaxies. Quasar fluctuations didn't seem so odd once it was determined that the nuclei of nearby galaxies can also undergo rapid luminosity changes.

Shortly after Schmidt's discovery, however, a small but very vocal group of astronomers began to argue that quasars were not located in the most remote corners of the universe. University of California astronomer Geoffrey Burbidge, Halton Arp of the Mount Wilson Observatory, and Fred Hoyle in Great Britain believed that the dim bluish specks were actually much closer to us. They were very uncomfortable with the idea that some mysterious machine a light-week wide or smaller was radiating hundreds of times more energy than an entire galaxy, which is a million times larger. This interpretation is not required if quasars are assumed to be objects residing nearby. By setting faint quasars closer in to us, their inherent magnitudes become rather nominal. Supporters of this notion suspected that the large redshifts in the quasar spectra were due, not to the expansion of the universe, but instead to some other bizarre process. The wavelengths of light coming out of a quasar, they contended, might be getting stretched as the photons struggle to escape some kind of intensely strong gravitational field; this would make the light appear "redder." On the other hand, claimed Arp, some entirely new principle in physics, as yet unknown, could be involved. More recently, proponents of this school of thought have suggested that there might be two classes of quasars: one type that is greatly redshifted by the universe's expansion; a second type that isn't.

Arp's arguments against very distant quasars weakened considerably as soon as several astronomical observing teams, working in such diverse locations as Hawaii, California, Arizona, and Chile, began to show that quasars are indeed associated with far-off galaxies. In 1979, for example, Susan Wyckoff of Arizona State University took particular interest in 3C 273, the quasar that started it all. She knew she had a difficult task ahead of her when she traveled to the Chilean Andes to use the 3.6-meter telescope at the European Southern Observatory. Studying a quasar is problematic because its relatively brilliant light simply overwhelms its surroundings. Trying to determine whether a quasar sits in the middle of a galaxy is much like trying to denote the size and shape of a burning match while a floodlight is shining in your face. "We certainly couldn't pull the photographic plate out of the fixer, hold it up to the light, and see right away that it was fuzzy, like a galaxy," says Wyckoff. Instead, she and her collaborator-husband Peter Wehinger had to first "digitize" the plate using a microdensitometer, a machine that scans a photograph and translates its regions of light and dark into a series of numbers that can be processed by a computer. "Once on computer," explains Wyckoff, "we simply subtracted out the pointlike quasar and immediately saw a dim nebulosity that was just the right size and luminosity for a galaxy at [the calculated] distance."

But was this merely a chance alignment, a galaxy caught right behind or in front of the quasar? The need to answer that question encouraged Wyckoff and Wehinger to return to Chile the following year to obtain a spectrum of their quasar "fuzz." It was a "blind" observation. "In the glare of the quasar," recalls Wyckoff, "we couldn't really see where the fuzz was. Instead, we had to place the slit of the spectrograph over a region where we thought it might be located. Then, we crossed our fingers and recorded the data for hours and hours." Not surprisingly, Wyckoff wanted to see the results before leaving South America. Since the European Southern Observatory didn't have a computer on hand, she and Wehinger traveled 100 miles along the Pan-American Highway to a neighboring observatory, the Cerro Tololo Observatory, to use its facilities. There, they worked for several days calibrating the data and converting the numbers into the proper units. Several spectra had been taken, but any one graph looked like a jumble of hash marks. "The moment of truth arrived when I could finally add all the spectra together," says Wyckoff. "It took just a few keystrokes, and when it was finally displayed, I'll never forget the sight. There before me was an emission line for oxygen with the exact same redshift as the quasar 3C 273—sixteen percent. At that moment, I knew that we had detected the gas of a galaxy, a galaxy that surrounded the quasar." Since then Wyckoff, Wehinger, and other researchers have found more than a hundred examples of quasar fuzz, strong evidence that quasars are truly the disturbed centers of otherwise average-looking galaxies. Quasars seem to be the most energetic specimens in a general class of active galaxies, which includes radio galaxies and Seyferts. Another common bond: The high resolution of radio-telescope networks has enabled astronomers to see that many "radio-loud" quasars, just like radio galaxies close by, channel much of their energy into two jets of plasma, which shoot out into space in opposite directions for hundreds of thousands of light-years. Astronomers are coming to understand that the universe offers a continuous spectrum of active beasts, from the relatively soft hum of a radio galaxy to the screeching wail of a quasar. What differs is the intensity of the outbursts.

"Ninety-nine percent of all stars are perfectly boring," points out Caltech astronomer Oke. "But we spend most of our time on the other one percent because their abnormalities reveal very important clues about stellar processes. The same is true for these rare active galaxies." Certain features of a galaxy, its hidden core, for instance, go on display whenever a Seyfert galaxy or a quasar undergoes its "pyrotechnics." It's a ready-made entree into unexplored territory. "The center," continues Oke, "is the one interesting physical point in any galaxy. That's where there's a discontinuity. If anything interesting or dramatic is going to occur in a galaxy, it's likely to happen in the center."

But what exactly transpires in that critical spot? What keeps a radio galaxy's cosmic river of plasma flowing or a quasar's blinding light shining? Most theorists agree that it must be a unique machine, whose design specifications are pretty stringent: For the most active galaxies, this mysterious engine has to be fairly stable (radio jets maintain their orientation for millions of years); extremely compact (daily and weekly fluctuations in galactic-core brightnesses suggest the engine may be as small as our solar system); and able to eject prodigious amounts of energy at nearly the speed of light, often channeling this energetic matter into two, oppositely directed beams. Researchers realized right away that thermonuclear burning, the simple fusing of atomic nuclei that drives every star in the universe, is not the answer at all. That would require squeezing a few galaxies' worth of stars into a solar-system-sized space, a very unlikely occurrence. But there are other possibilities. Astrophysicists are quite aware that the universe comes equipped with a much more efficient energy generator: When matter is thrown down a deep gravity well, the particles are accelerated to near-light speeds.

Gravity-driven motors, in fact, can generate up to one hundred times more energy than nuclear-fired engines. Supernovae and x-ray pulsars are vivid examples of this fact.

Presently, the most popular nominee for this gravitational pit, which allegedly dwells deep within the heart of an active galaxy, is a spinning black hole formed from the collapse of up to a few billion suns. Originally, these suns may have huddled together as an extraordinarily dense herd of stars, a type of cluster that could easily have developed in a crowded galactic center. It has been suggested that these stars, driven inward by the force of gravity, ultimately coalesced into a supermassive black hole. Some argue that a compact cluster of neutron stars or a supergiant pulsar could handle the job just as well; others contend that the formation of a black hole is inevitable in the end, a fate difficult to avoid in the evolution of a galactic nucleus. But since the existence of this stupendous creature is based only on circumstantial evidence, theorist Roger Blandford, a Britisher who has spent several years pondering this scenario from his Caltech office, discusses the model with a note of caution: "It's easy to get carried away with this idea, but you have to remember that we're drawing inferences from fragmentary circumstantial evidence. Not even by the most lax standards of scientific proof can we claim to have *demonstrated* the existence of a black hole within the nucleus of any galaxy." Yet, for the moment, the black-hole model appears to be the front-runner in explaining the peculiar outbursts of an active galaxy, be it a radio galaxy or quasar.

Blandford and others envision stars and gas in the active galactic center being sucked in by the powerful gravitational pull of the black hole, forming a doughnutlike ring, or accretion disk, that rotates around, and in the same direction as, the spinning hole. Enormous amounts of energy can be released as this maelstrom of matter spirals inward toward the black abyss and is ripped apart by the gravitational tug-of-war. If the hole is fed more than it can swallow, excess gas that has not yet reached the hole's infamous point of no return might be deflected—squeezed out like a cream filling from the top and bottom of the doughnut. This could be the source of those magnificent streams of matter so clearly imaged by the VLA. The radio signals would be broadcast as the charged particles underwent violent accelerations within the cosmic jets.

"But another very attractive way to supply the power," suggests Blandford, "is not to rely on the energy of the in-falling gas but rather to tap the spin energy of the hole itself. In effect, to create a cosmic dynamo." In this scenario, magnetic lines of force thread through the spinning hole's surface and whirl around with its swift rotation. Because of this tremendous spin, the magnetic field lines come out of the north and south poles of the hole coiled like streamers around a maypole, forming two narrow but powerful channels. Like a gigantic turbine in a cosmological power plant, these spinning fields would produce more than a million trillion volts of electrical potential, generating beams of particles that shoot out along each magnetic channel to near the speed of light. Again, such acceleration would trigger the particles to emit radio waves. "And the spinning black hole acts like a gyroscope," notes Blandford, "aligning the jets along its stable axis, which maintains a fixed direction in space."

So far, direct imaging of a galactic accretion disk has not been possible. If such a disk truly exists, its relatively small size makes it difficult for current instrumentation to resolve. But tantalizing clues have emerged to suggest that the swirling pool of matter is a real entity. At Lick Observatory, astronomers Joseph Miller and Robert Antonucci looked into the bright nuclei of two Seyfert galaxies and discovered that photons emerging from each core acted as if they had been scattered off a relatively small flattened object, perhaps an accretion disk. In 1980, Caltech astronomer Oke observed 3C 390.3, an elliptical galaxy with a particularly luminous nucleus located half a billion light-years away, and watched it "turn off" or at least drastically reduce the amount of energy it was emitting from its core. Over the following months, in direct response to that sudden drop, the radiations emanating from clouds of gas surrounding the galaxy's center also dimmed. The manner in which these clouds switched off strongly suggests that the gaseous material is arranged in the form of a disk.

But do these accretion disks necessarily surround a black hole? The evidence, so far, is only indirect. In the late 1970s, for example, American and British astronomers painstakingly measured the distribution of stars within M87, the supergigantic elliptical that sits in the very middle of the Virgo cluster of galaxies. They discovered that certain stellar motions within this enormous galaxy made no sense whatsoever unless they assumed that a dark, massive object inhabited M87's innermost regions. The mass of this enigmatic object is estimated to be at least five billion suns. Is it a black hole? Maybe. The event horizon of a supermassive black hole composed of five billion suns would occupy a region about the size of two solar systems. That would probably be one of the largest galactic black holes around, but "pint-size" versions may also form. Using the International Ultraviolet Explorer space telescope, a team of European astronomers studied the ultraviolet radiations being emitted by NGC 4151, one of the brightest Seyfert galaxies in the sky, and concluded that the nucleus of this spiral galaxy harbors a slimmer condensed object weighing a "mere" 100 million solar masses.

There are many ways that a galaxy can express its explosive personality, depending on the galaxy's type, its level of activity, and its alignment with Earth. By looking down the throat of a radio galaxy's jet, for instance, we see a BL Lacertae object or "blazer," a compact radio source that looks like a fuzzy blue star in visible light. Some active galaxies are particularly adept at spewing X rays (bright quasars can radiate nearly as much energy in X rays as visible light), while others are better emitters in the infrared portion of the electromagnetic spectrum. So, it's risky to classify the varied active galaxies, but there are two general categories: The first includes those objects whose outstanding feature is a resplendent core that blazes forth like a cosmic bonfire in the inky depths of space. Seyfert spiral galaxies and radio-quiet quasars (which may or may not be spirals) fit that description. The second includes those objects which exhibit prominent radio jets and lobes. Radio galaxies, most of which are ellipticals, and radio-loud quasars are in this category.

Black-hole devotees think these differences may simply reflect the environment surrounding the black hole; that is, the type of galaxy the engine resides in and the availability of fuel. Owing to their sluggish spins, or perhaps to mergers, ellipticals seem to form monumental central "stomachs." Yet many eat fairly delicately. "If you have just a little gas falling onto the hole [ellipticals are fairly gas-poor], then any radiation it makes can escape freely in radio jets. There's not much gas to get in the way," points out Blandford. Meanwhile, galaxies such as the Seyferts appear to assemble smaller stomachs, yet gobble up their vittles voraciously. "If a lot of gas is shoveled onto the hole," explains Blandford, "then the radiation can get trapped. The object ends up looking more like a star." Even if jets form, the gas-rich spiral disks may inhibit them from getting out.

Despite several decades of research on active galaxies and quasars, progress in understanding these exceptional beasts has been slower than many would have hoped. But then, muses Schmidt, this is not too surprising considering the obstacles astronomers face in this particular line of investigation. "Think of our neighbor, the Andromeda galaxy," he suggests. "Suppose it didn't appear to us as a spiral structure surrounded by globular clusters, but only as a point. How much knowledge could we extract from such a thing, compared to what we know now? Not much. That's what we're up against with quasars." Even as the Very Large Array was merely a sketch on an engineer's drawing board, radio astronomers knew that they needed interferometers with baselines of *thousands* of miles to delve farther and farther, into the very hearts of active galaxies, to see directly what was brewing. It's a matter of simple geometry: The smaller the source, the wider apart the radio antennas have to be to resolve the minute details. For a while, the situation looked hopeless, since a cable- or radio-linked system cannot easily span such transcontinental distances.

By the late 1960s, however, advances in atomic clocks provided the means to start pursuing that elusive galactic engine. At that time, groups in the United States and Canada began to simultaneously record the signals at widely separated radio telescopes on magnetic tape and to ship the tapes to a central computer, where they are combined to produce an image. The process can take weeks, for it involves processing trillions of bits of raw data. If observers fill up one tape every four hours and observe for twelve hours using six telescopes, they end up dealing with eighteen tapes. An extremely accurate atomic clock stationed at each antenna is the guarantee that the recordings (often done on modified video-cassette recorders) are synchronized to within a millionth of a second; any less precision produces a garbled picture. It wasn't hard to come up with a name for this new technique. If baselines of a few dozens miles are long, then antenna separations of several thousand miles are *very long.* Currently, Very Long Baseline Interferometry investigators regularly borrow time on half a dozen or more single dishes scattered about North America and Europe. At times, telescopes in South Africa,

Australia, and Japan are also brought into the network. Such an intercontinental array mimics the capability of an antenna thousands of miles in diameter—as wide as the Earth itself. In this way, Very Long Baseline Interferometry, or VLBI, attains angular resolutions 500 times keener than the VLA (and 1,000 times better than ground-based optical telescopes). This means that while the VLA and European arrays map the overall structure of radio galaxies, VLBI can discern details in galactic cores and quasars of only a few light-years' width—smaller than the separations between many stars in our galaxy. That's like being able to read a newspaper nearly 1,000 miles away—the distance from St. Louis to New York City.

While zooming in on the very fountainheads of some cosmic jets, VLBI has discovered some rather curious events. More than half a dozen quasars appear to be ejecting blobs of matter at velocities many times greater than the speed of light, in apparent violation of Einstein's theory of special relativity. From 1979 through 1981, for example, VLBI astronomer Stephen Unwin of Caltech and several colleagues used a network of five telescopes from California to West Germany to intermittently monitor a quasar known as 3C 345 situated more than eight billion light-years from Earth. Over that three-year period, they saw two knots break off from the quasar's bright core and race away at what appeared to be twelve times the speed of light. These knots traveled relentlessly onward, until they eventually faded away. Most likely, the blobs are not real lumps of matter, but rather shock waves moving down the quasar's jet like a celestial sonic boom. "There's actually nothing very strange about this so-called faster-than-light motion," Unwin assures us. "Einstein wasn't wrong. These superluminal movements are just not intuitive." Theorists contend that the faster-than-light, or superluminal, blobs are an optical illusion, a mirage that occurs when quasar jets are pointed almost directly toward Earth and the material streams out at velocities very close to the speed of light. Such geometry, coupled with the effects of the near-light velocities, makes the knots appear to fly outward at speeds much greater than they really are, a process that theorist Rees predicted would occur even before the first superluminal quasar was sighted in 1970.

As more and more dishes are brought into the VLBI network and better radio receivers are attached to each telescope, astronomers are able to peer into the fine structure of weaker, less flamboyant radio galaxies. They are finding that the cores of these mild-mannered ellipticals look very much like the nuclei of more active galaxies and quasars. Perhaps nearby radio galaxies are the glowing embers of quasars in their senior years. Conversely, they may be entirely different fellows—distant cousins that followed their own evolutionary path. The weak emitters might have come equipped with the same kind of engine as the 3C 273's in our universe, only one with a little less horsepower. The strange affairs going on in the middle of our own galaxy, spiraling streams of gas and hints of a smallish black hole, encourage the belief that this engine, although in a more dormant form, exists in the center of most galaxies.

Astronomers cannot hope to observe an evolutionary change in any one galaxy that will help them determine the exact connection between the active galaxies of yesteryear and the active galaxies of today. A human lifetime is but a momentary blip in the cosmological scheme of things. A galaxy changes appreciably only over millions, even billions of years. But there is a means of tracing evolutionary pathways in astronomy. Observers can look at particular populations of galaxies at successive distances and note the changes from one epoch to the next, similar to the way paleontologists sift through terrestrial strata to ferret out humankind's ancestral roots.

In 1972, for example, Maarten Schmidt and Richard Green, in order to discern the evolution of quasars, decided to conduct one of astronomy's most ambitious "digs" through the strata of space-time. Their strategy: photograph the northern sky, pick out from this field of six million stars the objects that are particularly blue, and record the spectra of the bluish spots to distinguish quasars from regular stars. The quasar's redshift would also peg its distance, enabling Green and Schmidt to see how the various quasars lined up in both space and time. After ten years of sifting through more than 2,000 bluish candidates (most of them turning out to be interesting but ordinary stars), they had their answer. The 114 bona fide quasars that they finally tracked down confirmed a trend noted earlier by Schmidt and others: As astronomers gaze farther and farther outward, they count more and more quasars with very high luminosities. Just a few billion years ago, these were a hundred times more plentiful. At an even earlier epoch, they appear to have been a thousand times more abundant.

There are actually two ways to interpret this data. For two decades, it was widely assumed that there were simply many more energetic galaxies in the distant past, which have now run down. By the time life developed on Earth, the very brightest quasars were a dying breed. According to this model, they are now gone, possibly living out their old age as quiescent black holes in ordinary-looking galaxies. "This gives one the idea that the quasar phenomenon is short-lived in a galaxy, perhaps a few hundred million years," says Schmidt. "A lot were born in the early days, but then as time went on, fewer and fewer bright ones appeared. They're like galactic fireworks, which last for only a short moment." Some of that *Sturm und Drang* could possibly be the birth pangs of galaxy formation itself.

But additional quasar data gathered by galaxy surveyors Koo and Kron led to the conclusion that very active galaxies were *not* more numerous in the distant past, merely more luminous (and hence more noticeable). They contend that the overall population of distant quasars is comparable to—or perhaps even less than—the number of Seyfert galaxies surrounding us today. Perhaps today's Seyferts and radio galaxies were simply far more brilliant billions of years ago. "Or maybe," poses Green, "galaxies are like the general population of stars." The most massive galaxies, he speculates, could have assembled their black-hole engines first, causing the galactic cores to burst forth in a glorious display. But then, like high-mass stars, these inflamed cores died off quickly. In the meantime, galaxies with lower-powered engines, such as the Seyferts and radio galaxies, may have come onto the scene and just plodded along in a less sensational fashion. This particular scheme is just one of several possibilities astronomers are debating.

Whatever the evolutionary pathway, there's a limit to the progression outward. When observers attempt to find any type of quasar situated more than 12 billion light-years away, a time when the universe was only a fifth its present age, they see nothing—only a primordial darkness. The luminosity of the quasars isn't the problem. Distant quasars are so radiant that they should be seen far past the 12-billion-light-year signpost. Princeton theorist Ostriker has postulated that our view is obscured at that point. "Dust obscuration is a problem that has bedeviled astronomers many times," he says. "It was the reason we didn't realize, right away, that the Sun was located at the edge of the Milky Way." Weather reports here on Earth often make such statements as "Visibility, two miles." Maximum visibility in the universe may be 12 billion light-years. But Schmidt has always suspected that this conspicuous cutoff marks a special moment in the history of the universe—when quasars first turned on to illuminate the dark recesses of the cosmos.

Though the fires of the mightiest quasars are probably now extinguished, they leave behind an energetic legacy, since their light is just now reaching us. Their emissions are so powerful that they bathe the entire universe in a diffuse glow of X rays. It's as if the universe underwent a youthful rebellion, but has now settled down to a more sedate middle age. Why has the universe shut down its bright quasar assembly line? There's one possible reason: Eons ago, galactic centers had lots more "food" around. Currently, a typical galaxy has about ten percent of its mass tied up in the form of gas and dust, but during its infancy it had much, much more. A newborn quasar could have gobbled up the centrally located rations, like a penitent coming off a fast. The central engine needs to be fed only a few stars each year, or their equivalent in gas and dust, to produce the kind of energies seen in a typical quasar. But the food supply is finite; a black hole has the power to suck in material only within a certain distance. "Today in many galactic centers the gas gauge says 'empty,' and there's no gas station in sight," comments Green, who continues his quasar research with the Kitt Peak National Observatory.

But extra food can be smuggled in. Evidence is growing that encounters between galaxies once served, and still act, as a potent trigger for all levels of galactic activity. For instance, the moderate activity in the center of our own Milky Way may be a reaction to its close encounter with the Magellanic Clouds. The strong radio-emitting ellipticals Fornax A and Centaurus A likely started broadcasting after merging with some spiral galaxies. And in galaxy–galaxy interactions, gas clouds within each galaxy, formerly in stable orbits, could collide with more frequency, tending to send gas inward toward the drowsing monster sitting in the nucleus. As University of New Mexico radio astronomer Jack Burns likes to point out, "nuclear activity within a galaxy may be no more than cosmic indigestion in the aftermath of a hearty meal."

Every galaxy, in fact, may have the potential to become active. Astronomer William Keel surveyed the 100 brightest—but otherwise normal-looking—spiral galaxies in the northern sky while he was with

Lick Observatory and noticed that 10 percent showed faint symptoms of feverish behavior in their nuclei. A similar study by Caltech observers detected quasarlike features in a third of their sample. This supports the notion that many present-day galaxies may, in fact, be "dead" quasars. "It's as if these galaxies were Seyferts or quasars with the engine on idle," says Keel. "The monster is there, but it's already grabbed all the gas it can get. Maybe the Seyfert phenomenon is episodic, occurring whenever two galaxies 'bump' into one another. Then, when they run out of food, the effect simmers down." Seyfert galaxies do tend to have companions, suggesting that their nuclear flares are kindled for a short while by the interaction. The black hole alleged to sleep in our own galactic center might wake up more fully when the Milky Way gets within half a million light-years of the Andromeda galaxy, four times closer than their present separation.

In the early days of the universe, when galaxies were much closer together, the chances of their brushing past one another or even merging were even greater. Quasars may have flared up because they are neighborly creatures. After examining several quasar outposts, in fact, Green and his colleague H. K. C. Yee noticed that quasars tend to hang out with other galaxies. They found that quasars, on the average, dwell within dense groups of galaxies more often than the typical galaxy. This may be significant in understanding the switch that turned a quasar on. Already, several quasars have been photographed with streams of material coming off of them, which appear to be the debris resulting from skirmishes with nearby companions. And when Dutch radio astronomer Rogier Windhorst hunted down the optical counterparts to the very faintest radio sources in the sky, he discovered that the majority of the radio signals were emanating from interacting galaxies; he had been listening to the sounds of galactic tête-à-têtes.

Further insights into the nature of active galaxies may be forthcoming from the realm of radio astronomy as it improves the resolution of its instruments. With ground-based interferometers limited to the width of the Earth, more resolving power will be available once VLBI takes to space. A large antenna orbiting 10,000 miles above the Earth's surface, for example, could increase current VLBI resolutions three times over. In the meantime, trans- and intercontinental arrays continue to stalk the enigmatic engines by gathering shorter and shorter radio wavelengths.

"The shorter the wavelength," explains Caltech radio astronomer Marshall Cohen, one of VLBI's founding fathers, "the deeper and deeper one can probe the central cores. It's like using X rays to look inside a grapefruit to see how big the seeds are. Maybe, just maybe, if we can get down to wavelengths of three millimeters, we'll finally get a look at the black-hole region itself." Or perhaps some other undreamed-of beast.

This slice of the sky, with the Milky Way situated at the apex, displays the universe's foamy texture over a distance of some 450 million light-years. Galaxies (represented by the individual dots) appear to be arranged on the surfaces of nested bubbles, which surround huge regions of nothingness. *Courtesy of M. Geller, J. Huchra, and V. de Lapparent. Copyright © 1987 Smithsonian Astrophysical Observatory Image Processing Lab/Michael Kurtz.*

7

CELESTIAL TAPESTRY

> Like the fifteenth-century navigators, astronomers today are embarked on voyages of exploration, charting unknown regions. The aim of this adventure is to bring back not gold or spices or silks but something more valuable: a map of the universe that will tell of its origin, its texture, and its fate.
>
> —Robert Kirshner

The story that unfolds in early man's most primitive records—at first, crude notches carved into chunks of bone; later, wedge-shaped characters impressed onto slabs of clay to keep track of the Moon's cyclical phases—indicates that his intellectual curiosity was repeatedly focused on the nighttime sky. He searched for order within the jumble of tiny lights scattered over the heavens. In many cases, this was done for practical reasons. The zodiac, groups of stars named after legendary heroes or familiar animals and objects, may have been carefully devised not only to keep track of the seasons, but as a navigational aid, enabling seafaring civilizations as early as 3300 B.C. to navigate the Mediterranean sea.

Seeking patterns in nature is one of science's most basic enterprises. The origins of plate tectonics, a concept that completely transformed the science of geology, can be traced as far back as 1620 when the English philosopher Francis Bacon commented on the parallel contours in the shorelines of two continents. By the 1960s, extensive exploration of the Atlantic seafloor showed that the "jigsaw-puzzle fit" of the shorelines of western Africa with eastern South America was more than coincidence. The distinctive pattern helped geologists see that the Earth's crust is divided into some twenty plates which move about like rafts over the planet's surface, thrusting mountains upward and deepening ocean trenches.

Now, astronomy is experiencing a similar breakthrough as the celestial landscape is charted. As astronomers improve their abilities to determine the redshifts of faraway galaxies, a growing number of surveys are being conducted to discern the three-dimensional structure of the universe; in other words, to map the way in which galaxies congregate throughout space. And these searches have revealed an unexpected texture to the cosmos. At first, astronomers found that many galaxies and clusters of galaxies appeared to be strung out along lengthy curved chains separated by vast regions of nothingness, areas seemingly devoid of galaxies. But such strings may be mere stitches in a much more complicated celestial tapestry. More recently, evidence has suggested that the universe is better described as foamy: as frothy, in fact, as the ample head on a newly poured glass of beer. Galaxies may actually be congregating in such a way as to weave immense, bubblelike structures through the universe.

Many of these filamentary and bubbly collections of galaxies stretch over hundreds of millions of light-years (a small yet appreciable fraction of the size of our visible universe), and this poses a worrisome challenge to an assumption known as the cosmological principle. This is the long-standing notion that, on the very largest scales, the galaxies are generally distributed quite evenly, making the universe as smooth and homogeneous as a fast-food milkshake. But the existence of gigantic voids, each the interior of an immense spherical shell of galaxies, makes the universe, on smaller but ever-enlarging scales, look more like a Swiss cheese or sponge. Much the way plate tectonics offered geologists unique insights into the formation of ocean trenches and continental mountain ranges, this foamy texture may inform astronomers how the lumpy islands of stars called galaxies originated from the Big Bang. The bubbly network could serve as a roadmap through time, back to the era when galaxies were first condensing. "Just as the texture of an antique fabric is the result of the vanished loom that made it," Harvard University astronomer Robert Kirshner has noted, "the present texture of the universe is a result of physical processes operating in a distant and inaccessible epoch."

The idea that the universe might be comprised of a hierarchy of structures—stars within galaxies; galaxies within larger groupings—was voiced as far back as the eighteenth century. In his *Cosmological Letters,* the Alsatian philosopher and scientist Johann Heinrich Lambert spoke of systems within systems within systems, all controlled by gravitation. This was almost two hundred years before it was widely accepted that other galaxies even existed in the universe. By the 1930s, extensive observations by astronomers such as Fritz Zwicky had demonstrated that galaxies do tend to gather into small groups and even larger clusters. Galaxies are, in fact, very sociable creatures. Even though space-time is continually stretching, moving most galaxies away from one another, gravity is strong enough to keep close neighbors together and to actually draw them closer. Our own Milky Way is part of a small aggregation of galaxies known (rather uninspiringly) as the Local Group, whose twenty-some members reside in a region approximately four million light-years wide. One end is anchored by the Milky Way, surrounded by a bevy of dwarf galaxies; august Andromeda (the only spiral galaxy that can be seen with the unaided eye) dominates the other end. The very richest clusters, though, make the Local Group look quite puny. A cluster located 300 million light-years away in the direction of the constellation Coma Berenices contains thousands of galaxies.

Whether clusters of clusters existed was never seriously considered in the first half of this century. Speculations, similar to Lambert's, had been published, but the cosmological principle was deep-rooted. During the 1930s, Edwin Hubble photographed selected regions of the sky and from a galaxy's apparent brightness (the faintest galaxies were assumed to be the most distant) crudely gauged the distances to thousands of galaxies. He concluded that galaxies and clusters were distributed fairly uniformly throughout the universe. It took the perceptive eye and feisty disposition of French astronomer Gerard de Vaucouleurs, newly graduated from the Sorbonne, to alter this widely held belief. In the early 1950s, de Vaucouleurs traveled from Europe to the Mount Stromlo Observatory in Australia to perform a tedious yet very important chore: a revision of one of astronomy's bibles, the Shapley-Ames catalog of bright galaxies. While updating the catalog's listing of southern galaxies, the young Frenchman couldn't help but notice that the Milky Way, along with its neighbors in the Local Group, was situated at the edge of a much larger system of galaxies. This system appeared to be a flat disk made up of multiple groups of galaxies. The Virgo cluster, a rich assemblage of several hundred galaxies located some 50 million light-years from us, serves as the system's centerpiece. Because we see this disk edge-on (just like the view of our own Milky Way), the many galaxies that comprise this system appear to form a long band that stretches across both the northern and southern skies. De Vaucouleurs was intrigued that the location of the Milky Way on the periphery of this disk was reminiscent of the outlying position of the Sun in our own galaxy.

De Vaucouleurs's finding was not an entirely new revelation. In the 1920s, Swedish astronomer Knut Lundmark, among others, reported that the bright spiral galaxies appeared "to crowd around a belt perpendicular to the Milky Way." But those who made mention of this remarkable clustering either were ignored by the astronomical community or failed to follow it up. De Vaucouleurs was more tenacious. He was the first astronomer willing to venture that this grouping was a distinct type of object in the universe. He dubbed it the "Local Supergalaxy," and thus revived Lambert's musings that the universe might consist of ever larger entities.

De Vaucouleurs, who joined the department of astronomy at the University of Texas at Austin in 1960, recalls that his suggestion "was received with resounding silence. It was considered as sheer speculation, even nonsense. Some prominent astronomers even told their students that it was an insane topic to work on. The concept that the universe was isotropic was too strong. It was dogma." Yet he waxes philosophic when looking back on those days of frustration. He likes to quote philosopher William James when summarizing the fate of such new discoveries: "First . . . it is attacked as absurd; then it is admitted to be true, but obvious and insignificant; finally, it is seen to be so important that its adversaries claim that they themselves discovered it." For de Vaucouleurs, it took more than a quarter of a century to be vindicated. Locating and mapping superclusters, cousins to our own Local Supercluster (as the Supergalaxy came to be known), has become a burgeoning endeavor in astronomy. De Vaucouleurs has written that "just as a growing child progressively becomes aware of larger units of human organization—family, neighborhood, city, and so on—astronomers have come in the past 400 years to recognize the hierarchical arrangement of the heavens. This astronomical growing-up is still in progress."

The study of large-scale structure in the universe was initially hindered by limited technology, astronomy's inability to determine, quickly and easily, the distance to remote galaxies so that a three-dimensional picture of the galactic distribution could be constructed. Although millions of galaxies can be seen on astronomy's most comprehensive map of the heavens—the 879 photographic plates from the National Geographic Society–Palomar Observatory sky survey—this sky atlas provides astronomers with just two of the three coordinates that locate a galaxy in space; it can pinpoint only a galaxy's "longitude" and "latitude," so to speak, on the surface of the sky. Pegging a galaxy's depth in space takes a little more work. Earlier, we saw how Hubble and Humason discovered the essential means of obtaining that third coordinate: The more distant a galaxy, the faster it recedes from us, shifting its light farther and farther toward the red end of the electromagnetic spectrum. A galaxy's redshift is directly proportional to its distance.

In Hubble's day, gauging the redshift for a galaxy, hence its distance, was a long and tedious task. An entire night was needed to record the spectrum of just *one* galaxy; sometimes, for the very faintest galaxies, the photographic plate had to be exposed over several nights. Humason, a consummate observer who started out as a mule driver for the Mount Wilson Observatory, would sit for hours looking at a guide star through the eyepiece of the telescope to make sure that the galaxy's image never strayed off the narrow slit of his spectrograph.

Today, except for some of the most distant galaxies, astronomers can duplicate Humason's work in a matter of twenty minutes, having replaced the inefficient photographic plate with state-of-the-art equipment such as image intensifiers and photon-counting devices. Sophisticated photon detectors can register anywhere from 20 to 70 percent of the impinging photons (compared to 2 percent or less for photographic plates). The highest efficiencies are achieved with an ultrathin, postage-sized wafer of silicon. The delicate chip, known as a charge-coupled device, or CCD, is no less than an electronic piece of film (invented at Bell Labs for picture-phones).

Photons of light entering the telescope strike the chip's surface and are converted into electrons that get stored in electronic wells at the point of impact. There are hundreds of thousands of these electronic wells, forming a gridlike pattern over the tiny CCD. These wells are essentially picture elements, much like the dots that make up a newspaper photograph. At the end of the exposure, the amount of charge stored in each well is read by a computer, which can then display the digital information on a graphics terminal as an image, an array of light and dark pixels. "The capabilities of these new devices have led to an explosion of data," says Margaret Geller, an astronomer with the Harvard-Smithsonian Center for Astrophysics who regularly analyzes redshift data to map the distribution of galaxies in the universe. "Because of these sensitive detectors, acquiring spectroscopic information has become routine. In the early 1970s, when I was a graduate student at Princeton, there were only a few hundred redshifts available for the nearby region. Today, we have tens of thousands." Indeed, this advance in technology has turned the study of large-scale structure, once the specialty of only a handful of observers, into a thriving branch of astronomy.

Some redshift surveyors began to assess the extragalactic "lay of the land" by conducting broad sweeps over large slices of the sky, a sort of astronomical strip-mining operation. One of the most extensive of these surveys was carried out by a team of astronomers from the Center for Astrophysics (CfA). Team member Marc Davis vividly remembers the struggle they faced to get the project off the ground: "The first two years were spent just redoing the instrumentation at our site on Mount Hopkins in Arizona. We used a 60-inch telescope there, and its instrumentation was very old-fashioned. If we'd had to deal with photographic plates, we'd never have been able to reduce the data. Photographic techniques are just not sensitive enough to do the work rapidly or accurately. We had to completely rebuild the spectrograph, assemble a modern detector, install a computer system, and write the operating software. For those first few years, I was spending so much time just trying to make the detector work that half of the astronomy faculty at Harvard were ready to kick me out." By 1978, though, the CfA system atop Mount Hopkins was off and running.

For some two hundred nights over a period of three years, the 60-inch reflecting telescope pegged the redshifts of some 2,000 relatively bright galaxies* across a third of the celestial sphere. The results were a 300-million-light-year panorama of extragalactic space, a depth five times farther out than our Local Supercluster. In their official report on the survey, the CfA team, led by both Davis (who went on to the University of California at Berkeley) and John Huchra, described the distribution of galaxies as filamentary "patterns of connectedness surrounding empty holes on the sky." Davis admits that, at the time, they were quite struck by the spongy architecture. "It really grabbed us," he says.

Additional features in the universe's structure were revealed as observers deeply probed smaller regions of the sky, areas that are particularly rich in clusters of galaxies. The idea has been that by plumbing a few deep samples, astronomers get a fair picture of what's happening all over the sky. Some have compared these efforts to a geologist's taking a core sample. Stephen Gregory, a pioneering redshift surveyor now with the University of New Mexico, has been sorting out the topography of rich clusters ever since his graduate-school days in the early 1970s. Over these years, Gregory and his associates Laird Thompson and William Tifft have trekked many times to Kitt Peak in southern Arizona to map particular patches of the sky. They have worked with a variety of telescopes located atop the mountain, but most often use the University of Arizona's 90-inch reflector. To pinpoint a galactic position, the team carefully aims the telescope at the target galaxy, sends its faint traces of light through a spectrograph, and records the data on a computer for a later distance determination. About every half hour, the

*Bright for a galaxy in the sky, but not to our eyes. The CfA survey looked at galaxies down to the 14th magnitude—galaxies about 3,000 times too faint to be visible to the naked eye.

procedure is repeated on yet another galaxy, until the arrival of daybreak.

There are no "Eurekas!" shouted out at the end of a night's work; the serendipity that pops up so often in other areas of astronomical investigation does not usually knock at this door. Most discoveries in large-scale structure come only after months, sometimes even years, of tracing the intergalactic byways. "There's an incredible amount of drudge work in this field," admits Gregory. "Doing the same procedure time after time after time." But the extraordinary patience of these redshift surveyors led to the awareness that the Local Supercluster, instead of being unique, is part of a larger celestial family. Among its members are:

- COMA SUPERCLUSTER The densely populated Coma cluster, with its thousands of galaxies, has long been a famous landmark to extragalactic astronomers. Redshift surveys suggest that a bridgelike connection joins Coma with another cluster called A 1367 (cluster No. 1367 in Abell's catalog of rich clusters). "The entire supercluster forms a long, beaded string that stretches 100 million light-years from end to end," says Gregory. Moreover, there seems to be a vast region devoid of galaxies between the Local Supercluster and Coma.
- HERCULES SUPERCLUSTER Astronomer Fritz Zwicky once called the assemblage of galaxies, situated in the direction of the constellation Hercules, "the largest conglomeration of matter so far known to us." There, several clusters combine to form a band, whose light takes some 500 million years to reach us. And, like Coma, it seems to have a vast empty region in front of it.
- PERSEUS–PISCES SUPERCLUSTER In the Perseus region of the sky, the A 426, A 347, A 262, NGC 383, and NGC 507 clusters line up to form a gently curving filament that Estonian astronomers Mihkel Joeveer and Jaan Einasto once described as a cucumber. At first, several groups working with optical telescopes confirmed that the Perseus-Pisces Supercluster, which lies more than 200 million light-years from Earth, spans 40 degrees in the sky, the width of eighty moons set side by side. But by using the world's largest single radio telescope, a 1,000-foot-wide dish that nestles within a natural bowl in the hills of Puerto Rico near the town of Arecibo, radio astronomers Riccardo Giovanelli and

Martha Haynes of the National Astronomy and Ionosphere Center at Cornell University were able to see that this filament is at least three times as long, maybe even more.

"We got into this field totally by accident," remarks Haynes. "We were observing a cluster in the Pisces sector of the sky with the Arecibo telescope to see how gas in spiral galaxies might be swept out of the disks as the galaxies moved about in the cluster. With that in mind, we needed to identify the edge of the cluster, but we couldn't find the edge at all!" The clusters in this region, they came to realize, were connected. Just as you tune your radio at home, Giovanelli and Haynes tuned the telescope so it could receive the particular radio signal, electromagnetic waves 21 centimeters long, given off by the cold hydrogen gas in the galaxies. Because these waves can travel with ease through the dust (the optical astronomer's bane) that lies along the plane of the Milky Way, the Arecibo team saw that the filament extends more than a hundred degrees across the sky. The Perseus-Pisces chain might even connect to another supercluster, a lesser-known group called the Lynx–Ursa Major Supercluster. And once again, voids were sighted nearby; the regions directly behind and in front of the Perseus-Pisces chain appear to be empty.

• PERSEUS-PEGASUS FILAMENT This is the granddaddy of them all. Astronomers David Batuski and Jack Burns of the University of New Mexico discerned this gigantic complex after running the Abell catalog through a computer and searching for clear-cut groupings. What they found was a snaky-looking filament, composed of twenty richly populated clusters of galaxies that stretches a *billion* light-years from end to end.

Superclusters are the largest collections of matter known to exist in nature. Each one contains the mass of at least 1,000 trillion Suns. But many astronomers confess that they were less surprised by the superclusters than by the huge pockets of what appears to be galaxyless space. Each empty region is the cosmic equivalent of the Sahara desert. To astronomers, it was like walking down the crowded streets of downtown Manhattan and suddenly coming upon several square blocks devoid of skyscrapers and brownstones. When Gregory and his colleagues first sighted a void in front of Coma, they were sure it was just a quirk;

they doubted they would observe many more. But supercluster hunters have now found voids scattered throughout the detectable regions of intergalactic space.

But many of these voids paled in comparison to the first appreciably large "hole" discovered in space—a gargantuan sphere of nothingness that researchers stumbled upon unexpectedly. In the late 1970s Robert Kirshner, then with the University of Michigan, and his colleagues, Augustus Oemler of Yale and Paul Schechter of the Mount Wilson Observatory, were probing a 35-degree patch of sky in the vicinity of the Boötes constellation in order to get a better idea of the average density of the universe. How many galaxies can you expect to find, they were asking, per unit volume of space? "We thought Boötes was a nice, boring section that represented a typical piece of the universe," recalls Kirshner. "Discerning the large-scale structure of the universe was not our intention, absolutely not."

But while they were counting the galaxies outward, using telescopes in Arizona and California, the Boötes probers realized that an unusual pattern was emerging. "Nearby, the redshifts indicated a high density, reflecting the presence of the Local Supercluster," says Kirshner. "But then, as we proceeded farther out, the density of galaxies went down dramatically. This was rather annoying since it made our analysis more difficult. We were hoping to avoid unusual regions of the universe."

Their curiosity piqued, the group, which now included Stephen Shectman of Mount Wilson, sampled the peculiar region more meticulously. "We eventually probed 282 very small fields over that entire sector, each only 15 arc minutes wide [the width of half the Moon]. It was like stuffing 282 knitting needles into a pumpkin to see if it has a hole in the middle," quips Kirshner.

Their exactitude paid off. They discovered that some 500 million light-years out from the Milky Way, in the direction of Boötes, there is a monstrous chunk of space with virtually no bright galaxies. The surveyors estimated that this cavity is spherical and spans nearly 300 million light-years, making it five times larger than previously sighted voids. Normally, a region this big would be expected to contain up to 5,000 galaxies; half the CfA survey would fit into it. Kirshner once jokingly called it the largest nonobject in the universe. "Of course, it doesn't have to be completely empty," he cautions. "It might be filled with extremely faint dwarf galaxies, or with gas that didn't coalesce." Both theorists and observers are seriously considering this possibility. But that, in itself, would be extraordinary. It would mean that the very early universe was somehow able to cook up different types of objects in different sectors of space.

Even before astronomers began mapping voids and lengthy superclusters, hints of a large-scale structure were there, owing largely to the extensive Shane-Wirtanen survey, which counted the myriad galaxies in the northern sky down to 19th magnitude (objects over a billion times fainter than the planet Venus when viewed as the morning or evening star). This undertaking was certainly one of the more remarkable projects in astronomy. Between 1947 and 1954, using Lick Observatory's 20-inch telescope, astronomers C. Donald Shane and Carl Wirtanen took 1,256 two-hour exposures—a set of photographs that covered the entire celestial sphere visible from their perch atop Mount Hamilton. Each photographic plate was 17 inches square and imaged an area of sky the size of the bowl in the Big Dipper.

Originally, this extensive photographic project was undertaken to study stellar motions. It was intended that a second set of photographs would be taken a few decades later to see how the stars in our solar neighborhood moved against what appears to be the fixed background of distant galaxies. But Shane, director of Lick Observatory at the time, didn't want to leave such a rich haul of data unattended. Shane noted in his autobiography that "in order to get some dividends from the program without having to wait twenty years for the second series of plates, it was decided, after consultation with Hubble, to use the plates for mapping the distribution of all the identifiable galaxies." And map he did. Over a period of twelve years, Shane and his assistant Wirtanen painstakingly counted each and every galaxy recorded on the plates. By photographing objects as faint as 19th magnitude, they were seeing the universe out to a distance of one billion light-years. Even before they completed half the counts, Shane and Wirtanen noticed that the distribution of clusters of galaxies appeared to be far from random. "The association of clusters to form larger aggregations or clouds of nebulae seems to be a rather general feature of the distribution. . . . One is tempted to speculate that clustering may be a predominant characteristic of nebular distribution," they reported. Their speculation later gained acceptance. Within two decades, those "clouds of nebulae" came to be widely known as superclusters.

By 1972, noted cosmologist and astrophysicist P. James E. Peebles of Princeton University turned to the Shane-Wirtanen survey to help him describe, in mathematical terms, just how clumpy the universe really is. In science, it's one thing to say, as Shane did, that something *looks* clumpy; it's another thing to prove it. Several researchers had attempted the proof as far back as the 1930s, but they lacked both the data and the computational machinery. "We came along," says Peebles, "when both were in place. These things are never a bolt out of the blue." Throughout the seventies, he and a number of associates, many of them graduate students working on dissertations, verified that galaxies are not just randomly scattered over the sky, but congregate in ways that can be mathematically defined. They were no less than shrewd gamblers, whose turf was the universe instead of Las Vegas. Given the position of one galaxy, they were tallying the odds of finding another galaxy at some specified distance. Their equations were the first comprehensive handle on the amount and extent of the universe's clumpiness.

This is the type of research that rarely receives much attention outside of a select circle of specialists, but the Princeton effort was an exception. While trudging through their statistics, Michael Seldner, Bernard Siebers, Edward Groth, and Peebles plotted the Shane-Wirtanen galaxy counts onto a two-dimensional map of the celestial sky. Their grid was composed of roughly one million pieces, and they made each bit a varying shade of gray (ranging from black for zero galaxies to white for ten or more galaxies) depending on the number of galaxies that Shane and Wirtanen counted in that pixel. The result was so eye-catching that it was made into a popular poster entitled "One Million Galaxies," which can probably be found in just about every college astronomy department in the United States.

Astronomers and laymen alike have gazed at this fascinating picture and commented on the immensity of space-time that it imparts. Yet it should be remembered that the Shane-Wirtanen survey map portrays only counts of galaxies; no real distance information is provided. Galaxies near and far are piled up on one another at each point. Still, it reinforced a growing conviction among astronomers that the universe over distances of many tens of millions of light-years was indeed filled with a network of knotlike galaxy clusters, curving filaments, and empty voids.

Astronomers had to be very careful, though, in arriving at that conclusion. The eye is very sensitive to patterns, a trait that enabled humans and their predecessors to spot a predator amid the jungle foliage. Psychologists have proved that we are quite capable of finding elaborate features in haphazard scatterings of dots. For a long time, many astronomers were not convinced that lengthy chains of galaxies and clusters, stretching over hundreds of millions of light-years, were anything more than statistical flukes, the eye picking out patterns in a randomly clumpy distribution of objects. Astronomers, in fact, have often fallen victim to nature's mischievousness. In the opening decades of this century, for instance, spirals were the rage. Astronomers saw that most nebulae in the nighttime sky tended to be spiral (before they realized they were actually "island universes"), so it was natural to search for spiral patterns in other astronomical objects. At least one journal paper was written describing a four-armed spiral pattern in a globular cluster, which is a perfectly spherical ball of stars. "The eye was similarly deceived in the case of canals on Mars," points out Peebles. But as evidence mounted that voids and superclusters were bona fide structures, researchers became encouraged that the topography could shed light on one of astronomy's outstanding mysteries—how galaxies came to be born.

To many redshift surveyors, the discovery of large superclusters and giant voids strongly supported a model of galaxy formation that came to be known as the "top-down" model since it envisions galaxies fragmenting from larger, primordial structures. This is a theory that originated in the Soviet Union in the early 1970s. Theorist Yakov Zel'dovich of the Institute of Applied Mathematics in Moscow and several colleagues argued that flat, gargantuan clouds of gas were the first entities to emerge right after the Big Bang. Each of these "pancakes," as the Soviets dubbed them, would have stretched over tens of millions of light-years and contained enough material to form some 1,000 trillion suns. Galaxies came about, they contend, as turbulence and shock waves fragmented the massive sheets into smaller pieces. The voids and filaments would appear quite naturally as the randomly oriented pancakes grew and intersected over time. It was even suspected that the superclusters might connect to form cells, weaving a pattern (as one astronomer put it) resembling a lace tablecloth.

Recreating the development of galaxies in the universe has never been an easy exercise for theorists, for the Big Bang generated a remarkably smooth ocean of plasma. Cosmologists know this because a remnant of the original fireball, a faint background of microwave radiation that permeates the entire universe, is incredibly uniform. In order to create the lumps that eventually developed into galaxies, most theories of galaxy formation start with the premise that the primeval blast somehow sent a series of waves rippling through the newly born sea of particles, both large and small fluctuations in the density of the gas. To Zel'dovich, fast-moving photons in the early universe would have completely damped the "tiny" galaxy-size ripples, giving the large-scale undulations free rein. As these monstrous waves moved through the cosmos, they allegedly broke the smooth, primordial ocean of matter into supercluster-sized segments.

"Top-downers" have been quick to point to several pieces of evidence in their favor. University of Hawaii astronomer R. Brent Tully, who conducted a nine-year radio survey of the Local Supercluster's structure, reported in the *Astrophysical Journal* in 1982 that "the thinness of the disk of the supercluster, the extreme segregation of the galaxies into a small fraction of the volume available, and the low local random motions are all evidence . . . in favor of the viewpoint that galaxies fragmented out of larger scale structure." Tully and his collaborator J. Richard Fisher of the National Radio Astronomy Observatory measured the width of the Local Supercluster disk to be six times greater than its thickness, making it look suspiciously like a pancake. Later, Tully came to realize that this was just the tip of the iceberg. He has gathered evidence suggesting that the Local Supercluster is just a small part of a much larger structure: a flattened system, composed of massive parallel sheets, that encompasses the Coma, Perseus-Pisces, and Pisces-Cetus Superclusters. "It seems that there is a propensity in the universe for things to lie in flattened planes," says Tully.

But the verdict at that time was far from in. When it comes to explaining how galaxies formed, the top-down theory is a relative newcomer. Its major competitor has been an older, more conservative hypothesis called hierarchical clustering, a process which is the exact opposite of the top-down scenario. This idea has long been championed by Princeton's Peebles, who believed that galaxies formed first, then gravitationally gathered into clusters and later superclusters as the universe evolved. It's been tagged, appropriately enough, the "bottom-up" model. Given the laws of gravitation, it has always seemed more logical to "bottom-uppers" that smaller knots of matter, pushed and squeezed by those ripples emanating from the Big Bang, would be the first to coalesce.

Supporters of the bottom-up school of thought are convinced that today's superclusters are not the remnants of primordial pancakes. To the contrary, they're sure that superclusters are rather young. The Milky Way, for instance, is not moving away from the Virgo cluster as fast as it should with the universe's expansion. This is a sign that the Milky Way is being drawn in toward Virgo, the center of the Local Supercluster. "To me," says Peebles, "this is a symptom of a supercluster in the process of formation. If the Local Supercluster had emerged before the galaxies formed, then today it would have been five times more compact."

Meanwhile, space x-ray telescopes, including the *Einstein* observatory, have revealed that clusters of galaxies are other than what astronomers long assumed them to be. In many clusters, x-ray-emitting gas does not shroud the entire assembly like a smooth, unwrinkled blanket. Instead, there are pockets of gas, each marking off a separate chunk of the cluster. This seems to indicate that the clusters are not old, settled, and thoroughly mixed (as once thought), but rather young and vibrant. Other studies support this conclusion. The Cancer cluster, for example, was found by optical astronomers to be a collection of five small discrete groups that are just now coalescing. "The prime example of this phenomenon is the Virgo cluster," says Harvard's Huchra, who has spent many years "dissecting" the cluster's structure. "A set of elliptical galaxies are settled in the core, while a number of spiral galaxies are still falling inward at several hundred miles per second. The entire cluster appears to have formed only in the last third of the universe's history, the last five billion years."

And the voids? According to bottom-uppers, a cluster can act very much like a gravitational vacuum cleaner. As the galaxies begin to congregate, they leave behind large sections of cleared-out space. Hence, it might not be so surprising to find big holes next to clusters and superclusters. Such segregation could all be part and parcel of the dynamics of clustering (though disciples of the hierarchical process will admit that sweeping out voids as large as the Boötes hole is still a problem for them).

The fact that the top-down and bottom-up theories each have their merits is perhaps a reflection that the overall topography of the universe has not yet been mapped. "As cartographers of the universe, we're about in the same position that the Greeks were 2,500 years ago," says Geller. The Greeks knew the Earth was spherical and could estimate its circumference with astounding accuracy. But they had not an inkling of the true extent of the continents and oceans. "In absolute size," continues Geller, "the volume of space covered by the original CfA survey is huge, about ten trillion trillion cubic light-years. But that's still only one hundred-thousandth the volume of the visible universe. A map of one hundred-thousandth of the surface of the Earth would cover an area only a little bigger than the state of Rhode Island. I'd hate to extrapolate the entire geography of Earth from it. Would we be able to guess at the existence of the Himalayas or the Sahara desert by looking only at Rhode Island?" Probably not. But ever so slowly, the borders of the celestial map are being extended.

During the winter and spring of 1985, astronomers from the Harvard-Smithsonian Center for Astrophysics proceeded to extend the original CfA redshift survey, probing nearly two times farther into space. Guided by John Huchra, a team of observers pegged the locations of hundreds of additional galaxies in a wedge of sky 117 degrees wide and six degrees thick. "Eventually," says Huchra, "we'll determine the redshifts of twelve thousand galaxies. The bigger the volume, the better chance we have at determining the true structure of the universe."

When Valerie de Lapparent, a University of Paris student studying under Geller and Huchra, plotted the first extended data as part of her doctoral thesis, she saw that the galaxies congregated as if they were the surfaces of huge, nested bubbles. Every void was the interior of a roughly spherical shell of galaxies. The flattened association of galaxies to which the Milky Way belongs, the Local Supercluster, could then be interpreted as being part of a bubble surface. The cosmos was not entirely filamentary; it more nearly resembled a kitchen sink full of soapsuds.

The new data led de Lapparent, Geller, and Huchra to conclude that some of the strings of galaxies sighted by earlier redshift surveyors may be illusions, merely slices out of the larger shell-like assemblies of galaxies. It was as if they had been charting the thin crust around one slice of bread, when in fact the crust was more extended and enclosed an entire loaf—a void.

Both large and small voids are present in the revised CfA map. The largest stretches some 240 million light-years across, comparable to the Boötes void. The Boötes void was once unique; now, such large voids appear to be quite common. Moreover, the Coma cluster was found to sit right at the intersection of several bubbles. This finding could shed new light on the formation of rich, dense clusters.

The sharply defined surfaces of the bubbles and the emptiness within suggest that gravitational forces alone did not construct the foamy structure; gravity could not have held such cells together over time. The extragalactic suds, in fact, look suspiciously like superbubbles, the spherical regions of space in our own galaxy that have been swept clean by a series of supernova explosions. Geller has pondered whether a similar hydrodynamic process—albeit on a much bigger scale—could have generated a bubbly structure within the universe. An alternate theory of galaxy formation, first proposed in 1981, is actually based on this idea.

Princeton astrophysicist Jeremiah Ostriker and Lennox Cowie of Johns Hopkins University have suggested that the formation of stars and galaxies themselves could serve as powerful generators of voids and shells of galaxies. "Right after the universe's birth, everything was hydrogen gas. It was very, very uniform. But now when we look around us it's very lumpy. How did it get from one state to the other?" ponders Ostriker. "You know the old saying, 'You can't get there from here.' Well, in this case, it's the other way around. We can't seem to get here from there. How do you forge a lumpy universe from a formerly smooth one?" Ostriker often reminds his colleagues that gravity and perturbations from the Big Bang are not the only game in town when it comes to forging large-scale structures. Instead, he imagines the early universe giving birth, fairly quickly, to massive stars that raced through their nuclear cycles, releasing tremendous amounts of energy as they burned and explosively died. There are hints of this having happened. The oldest stars so far observed in our galaxy are hardly pristine; they contain appreciable amounts of heavy elements, such as carbon, oxygen, and calcium, in their outer envelopes. Such "contamination" must have come from an earlier generation of stars, a large portion of them dying and scattering their ashes before galaxies fully coalesced. "The question is," says Ostriker, "what happened to that energy? It's likely to have sent a shock wave propagating outward in all directions like a bomb explosion. Once galaxy formation gets started, it can drive itself, just as star formation induces more star formation in our own galaxy."

Like a snowplow, this cosmic shock wave would sweep any matter in its path into a thin shell, which fragments into new stars and galaxies; from there, the process would continue ever more rapidly. After several cycles of this "explosive-amplification" process, as Ostriker and Cowie call it, the galaxies would end up arranged in flattened superclusters (the remnants of those thin, fragmented shells) surrounded by vast regions of nothingness. At present, though, this mechanism still has trouble generating lots of voids as big as the Boötes "hole."

In order to confirm or refute these varied scenarios, astronomers will have to make further observations. "Unlike other sciences, we can't go into a laboratory and give these systems a kick to see what they do," Geller reminds us. "The only way we can understand the large-scale structure is to continue measuring lots of redshifts." Some answers may be forthcoming as astronomers extend their redshift surveys to ever wider and deeper regions of the sky to get a better look at the spongy texture. Are voids merely statistical quirks, astronomers want to know, or are they as prominent and numerous as the holes in a sieve? Space telescopes launched into earth orbit in coming years will be a valuable tool. Their mirrors and lenses, far above our planet's roiling atmosphere, will take astronomers back to a time when the universe was much younger, helping them see which came first: supercluster-sized structures or individual galaxies. More important will be the construction of the next generation of ground-based optical telescopes, which will also benefit from the latest advances in technology. The planned 10-meter Keck telescope, to be installed on Hawaii's Mauna Kea by Caltech and the University of California, will have a mirror with four times the light-gathering power of the 200-inch Hale telescope. Since a single mirror, nearly 400 inches across, would severely sag under its own weight, the Keck mirror will be constructed out of thirty-six hexagonal segments, each nearly 6 feet wide, kept in position by a computer. This advance will enable redshift surveyors to map the very farthest frontiers of the universe.

But the full story may not be revealed until astronomers unravel what has come to be known as cosmology's "dire secret": Most of the mass of the universe, perhaps more than 90 percent, cannot be directly seen by astronomers, not even with the most sensitive radio, infrared, or x-ray instrumentation. Over the last few decades, evidence has been mounting that an unidentified dark material pervades our cosmos. Some researchers are asking whether astronomers can really hope to understand the workings of the universe and its ultimate fate if the objects of their attention—exploding galaxies, spinning neutron stars, and dusty molecular clouds—are merely luminous flotsam within an ocean of invisible matter. It may be like trying to ascertain what a potato, buried deep in the ground, looks like by analyzing its leaves. As we shall see, deciding whether supercluster-sized clouds of gas, tens of millions of light-years across, fragmented to make galaxies or, instead, small, galaxy-sized lumps formed first and gravitationally flocked together over the eons may ultimately hinge on discovering the true nature of this pervasive, yet elusive, cosmic material.

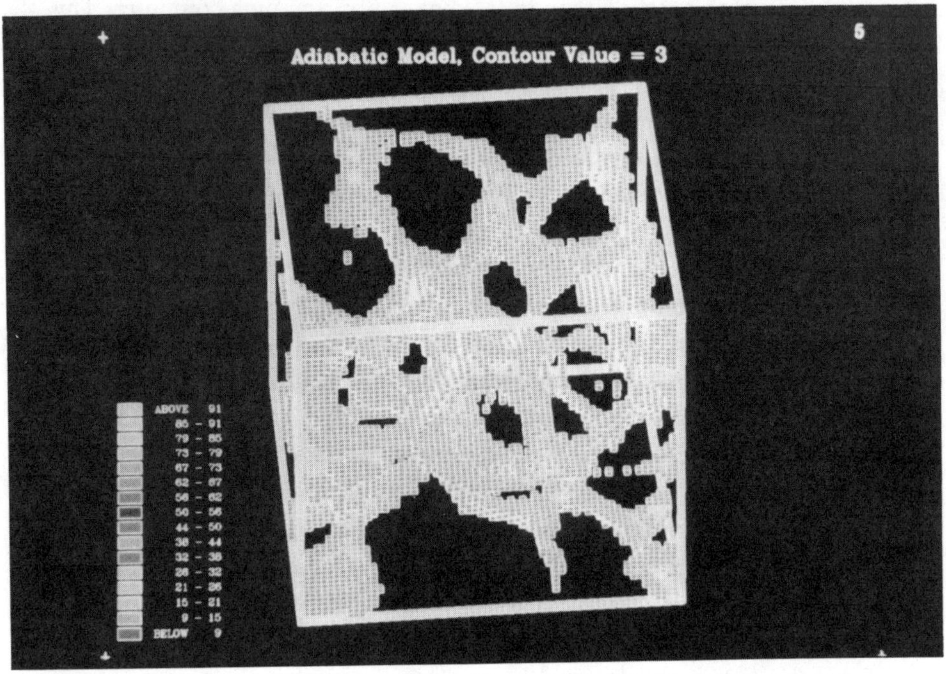

By assuming that massive neutrinos comprise 90 percent or more of the material in the cosmos, computer simulations of an evolving universe present a "Swiss cheese" structure: filamentary neutrino clouds surrounding large voids. *Courtesy of Joan Centrella.*

8

THROUGH A UNIVERSE, DARKLY

O dark dark dark. They all go into the dark,
The vacant interstellar spaces, the vacant into the vacant . . .

—T. S. Eliot, *Four Quartets*

As astronomers expanded the scope of their operations into varied regions of the electromagnetic spectrum, they learned a lesson more often associated with the inscrutable world of magic: What you see is not necessarily all that is there. Myriad protostars, which gestate within thick cocoons of dust and gas, remained out of sight until the advent of sensitive infrared instrumentation. Radio telescopes were needed to discern the rampaging streams of electrons flowing out of galactic cores. And the celestial landscape continues to be refashioned.

Recent studies of spiral galaxies, including our own Milky Way, indicate that the spiral disks rotate in an unexpected manner. Clusters of galaxies, too, display strange internal motions. This evidence suggests to astronomers that *something else* lurks out there in intergalactic space. They call it "dark matter," simply because this unknown material pervading our cosmos doesn't seem to absorb or emit any kind of radiation (at least, any x-ray, radio, or light waves that can be detected so far). The situation is intriguing, mysterious, and downright embarrassing. The bulk of the universe's mass could be playing an awfully good game of hide-and-seek with all our earthly detectors. "It's a puzzle that confuses us more and more," comments Princeton's Ostriker. "We've come to see that the luminous stars and galaxies, long described by astronomers, are only a few percent of what's actually out there. Most of the matter in the cosmos is invisible, and we don't have a clue as to its composition. If we let history be our guide, then our picture of the universe is likely to be significantly altered and enlarged as we attempt to solve this mystery."

Astronomers are baffled. The extent to which they are stumped can be measured by the astounding number of diverse objects put forth as candidates for this dark matter—from swarms of hypothetical elementary particles, each mote weighing less than a trillionth of a trillionth of a trillionth of a gram, to a hidden armada of Jupiterlike planets. "Anyone having a clue as to the whereabouts of this missing mass should contact the nearest observatory immediately," astrophysicist Wallace Tucker once wrote. "Reward for information leading to the discovery of missing mass: possible fame; no fortune."

The first glimmer that something was amiss in astronomy's understanding of the universe came in the 1930s. Caltech astronomer Fritz Zwicky, an eclectic wizard at his craft, was measuring the velocities of galaxies within the famous Coma cluster and noticed that they were moving at a fairly rapid pace. He added up all the light being emitted by these galaxies and realized that there was not enough visible, or luminous, matter around to gravitationally bind the speeding galaxies to one another. Under the standard laws of celestial mechanics, the Coma cluster should be flying apart, but it isn't. The situation seems paradoxical. It's as if space-shuttle astronauts ignited all the engines and rocket boosters on their spacecraft at full power, yet found themselves unable to lift off from the Earth. Shuttle engineers might be forced to conclude that the Earth suddenly and inexplicably had more mass, and consequently a stronger gravitational field to keep such things as shuttles from flying off too easily. Similarly, Zwicky had to assume that some kind of dark, unseen matter pervades the Coma cluster to provide an additional gravitational glue. This unseen ingredient came to be known as the "missing" mass, though, of course, it really isn't missing. The manner in which the galaxies move about one another reveals that something *is* there. What is missing is the light.

The problem of the missing mass didn't make much of a splash during those Depression years. Some astronomers thought it was just an exotic but not too bothersome property of clusters; others believed the dilemma would disappear once theorists figured out the motions of galaxies in more detail. "Weighing" a cluster, they were sure, had to be much more complicated than Zwicky had assumed. But their thoughts were colored by their prejudices. Except for a handful of farsighted visionaries, astronomers at that time were still wedded to the optical portion of the electromagnetic spectrum. With the establishment of radio, infrared, and x-ray astronomy yet to come, many observers were not comfortable with the idea that much of the universe's activity is invisible to our eyes.

By the 1970s, however, the problem of the missing mass was brought closer to home. By then, both radio and optical telescopes were beginning to reveal curious rotations in both the Milky Way and nearby galaxies, which suggested that galaxies contained more mass than previously assumed. Astronomers always took it as a matter of course that stars in a spiral galaxy would revolve around the galaxy's core like planets in a solar system, whose motions adhere to Newton's law of gravitation. Newton recognized that the gravitational attraction between a planet and the Sun follows a simple rule of thumb: The attraction between two celestial objects is inversely proportional to the square of the distance between them. That means that if the distance between the Earth and the Sun were doubled, their mutual gravitational grip would lessen by a factor of four. Triple the distance, and the attraction would fall off to a ninth of its original strength, and so on.

The distance between a planet and the Sun also determines the planet's orbital velocity. "In the solar system, the planets all orbit the Sun with velocities that get smaller and smaller as they get farther from the Sun, the system's center of mass," explains astronomer Vera Rubin of the Carnegie Institution of Washington. "So the inner planets go fast, and the outer planets go slow. That's just a direct response to Newton's law." For example, the planet closest to the Sun, Mercury, whips around the Sun at a racy 108,000 miles per hour. Faraway Pluto, on the other hand, orbits at a sedate 10,500 miles per hour; if it went any faster, it would fly out of the solar system. "It was always expected that the same thing would happen in galaxies," continues Rubin, "that the stars and gas clouds in the outer edges would be orbiting slower than the inner, more massive parts."

But to everyone's surprise, observers discovered that galaxies weren't acting like gigantic solar systems at all. Radio astronomers were among the first to notice. By tuning to hydrogen's special radio song, they observed that gas in the outer regions of a luminous spiral disk moves just as fast as the gas located in the inner galactic core, sometimes even faster. "This meant," says Rubin, "that the distribution of mass within a galaxy was not at all like the distribution of light," which is greatest in the central regions. Yet, for a long time, the galaxies' odd spins were either attributed to instrument error or merely ignored. They could not be swept under the rug, however, once Rubin obtained a veritable arsenal of evidence that reinforced the unexpected finding.

Vera Rubin has spent most of her professional life studying galactic rotations; her conclusion that something strange is going on within galaxies was not arrived at hastily. Starting in the mid-1970s, to confirm her suspicions, Rubin undertook one of her most ambitious projects. Over a number of years, using telescopes in Arizona and Chile, she and several colleagues measured the rotational velocities of more than seventy varied spiral galaxies: dim ones, bright ones, spirals with starry arms tightly wrapped, and others with arms splayed open. Specifically, they analyzed the light emanating from rich clouds of gas that surround hot, young stars in these spinning galaxies. The trick was to pick out galaxies whose disks were oriented edge-on to the Earth, or nearly so. In this way, the rotation of the galaxy carries the stars and gas on one side of the disk toward the Earth and those on the other side away from it. As a result, the wavelengths of light emitted by the gas clouds approaching the Earth get a bit shorter (that is, become "bluer"); the light waves lengthen (get "redder") for the gas moving away. The velocities of stars and gas across each spiral disk could then be pegged by carefully measuring the extent to which the wavelengths shift.

In spiral galaxy after spiral galaxy, the Carnegie group saw that stellar material on the outer edges of a disk travels around at speeds much faster than theory had estimated. It was the Coma-cluster problem all over again. It seemed that a significant reservoir of unseen matter had to be tucked away somewhere to keep the stars from zipping out of the galaxy. Otherwise, spiral galaxies would have been out-of-control merry-go-rounds that broke up soon after they were born. This interpretation, it should be noted, is not quite unanimously accepted. A few theorists have wondered whether these unexpected motions within galaxies and clusters are actually telling us that Newton's law of gravitation breaks down over distances of tens of thousands of light-years. But this radical suggestion has not been greeted enthusiastically. When faced with the necessity of accepting the existence of dark matter or modifying one of astronomy's theoretical cornerstones, most astronomers will opt for the former. "The conclusion is inescapable," says Rubin. "Mass, unlike luminosity, is not concentrated near the center of spiral galaxies. There has to be *lots* of mass farther out, beyond the visible disk. What our work has shown is that dark matter is present in individual galaxies, as well as in clusters."

Earth's home galaxy is no exception. From all the light being emitted by the Milky Way, astronomers have estimated that our visible galaxy contains the mass of about 100 billion suns. But gravity, which acts as a kind of cosmic weighing scale, tells another story. The Milky Way is being drawn toward its neighbor Andromeda as though our galaxy had a mass *ten* times greater than 100 billion solar masses. This is a frustrating experience for astronomers; it is as if they are hearing footsteps all around them, yet are unable to see the intruders. Where could all this extra mass be located, they asked?

In 1973, Princeton colleagues Ostriker and Peebles stumbled upon one possible hiding place while they were studying the structure of the Milky Way. "Jim and I made a numerical model of the disk," recalls Ostriker, "and put it into the computer. To our surprise, the disk went wildly unstable. The stars' orbits went from being nearly circular to being very eccentric. Some of the stars even flew off." The two theorists concluded that spiral galaxies couldn't retain their usual shape unless they were embedded in a vast sphere of material. Every glowing galactic disk, they conjectured, was immersed in a bloated halo of dark, invisible matter. At first, this concept was greatly resisted, but eventually observers latched onto the idea with relish, since it gave them a conceptual hook on which to hang their missing-mass hats. The rapid speeds at which the globular clusters and Magellanic Clouds move about the Milky Way, far above the galactic plane, suggest that the diameter of this halo could be, at the very least, five times wider than the visible disk of stars and gas. The clusters in those far regions are moving so fast, say some astronomers, that they have to be immersed in an extensive sea of unseen material to stay confined to the Milky Way.

Even though spiral galaxies have garnered most of the attention up to this point, elliptical galaxies are not exempt from possessing a missing mass. X-ray telescopes have revealed that M87, the giant elliptical that sits in the center of the Virgo cluster, is enveloped by a diffuse yet massive cloud of hot gas. Unless M87 contained much more matter than previously suspected, this gas, heated to temperatures of millions of degrees, would have escaped into intergalactic space long ago. What is required is a galaxy containing at least 30 trillion solar masses—the combined weight of 300 Milky Ways!

Many blame the missing-mass problem on sheer bookkeeping errors: neglecting to count all the very faint stars, cometary debris, dark planets, and gas inhabiting the heavens. Indeed, as astronomers improve the efficiencies and sensitivities of their detectors, they uncover new caches of material almost daily. While space telescopes such as the *Einstein* observatory observed clusters of galaxies immersed in large pools of x-ray-emitting gas, radio astronomers were detecting immense clouds of hydrogen in those seemingly empty gulfs between galaxies. University of Michigan astrophysicists Dennis Hegyi and Garth Gerber accounted for an additional fraction of material when they used a sensitive photon-gathering device to look at a typical spiral galaxy called NGC 4565. Instead of this galaxy appearing as a very thin disk, which is how most spiral galaxies appear in standard astronomical photographs, it actually looked a bit puffed up. A faint halo of stars makes the disk at least ten times thicker. And, lastly, the recent surge in planet-hunting around our solar neighborhood has certainly proved that extrasolar planets are more plentiful than once thought.

Although newfound x-ray-emitting gas, faint stars, and intergalactic clouds can as yet account for only a small fraction of the missing mass, a large number of astronomers are perfectly happy with the idea that more rigorous stalking will balance the books. More advanced x-ray space telescopes, for example, might spot diffuse X rays given off by a galactic halo populated with dim red-dwarf stars. Astronomers who lean toward this view are spurred on by a fascinating concurrence. At this time, the velocities of galaxies within clusters and the unusual rotations of spiral galaxies suggest that there is about ten times more dark stuff in the universe than luminous matter. Taking this dark and luminous matter as a whole, that just about matches the maximum quantity of protons and neutrons, the two major constituents of atomic nuclei, estimated to have been forged in the first second of the Big Bang. Collectively known as baryons among physicists, these protons and neutrons basically comprise everything we see around us, from people to planets. "We hope we have the physics right," says Peebles, who has dealt with these calculations. "But isn't it the most daring extrapolation you can imagine—to say that the universe behaved, at the ripe old age of one second, exactly as we compute it 15 billion years later? Maybe it's just coincidence."

Many astronomers believe that it is a very telling match-up—strong evidence that the missing mass can simply be made out of familiar but not-easily-detectable astrophysical orbs. Others argue, however, that there are good reasons to suspect that the bulk of the dark matter is not composed of "ordinary" baryons. Objects made of baryons are fairly noisy creatures; they love to advertise their presence by moving about, radiating, or even blowing up. If the spherical halos of dark matter were composed of cold gas, for example, the unenergetic atoms would have fallen into the centers of galaxies eons ago, and a hot gas would be radiating more X rays than astronomers now observe. Aged stars, such as tiny white dwarfs and neutron stars, are also poor dark-matter candidates. The death of a star is an operatic event whose crescendo scatters lumps of matter far and wide. Some astronomers believe that if the dark matter exists in the form of dead stars, they would be detecting the leftovers by now.

There is one loophole in this argument: The explosive deaths may have occurred far enough in the past that the reverberations died out long ago. "The pregalactic universe may have been full of massive stars forming at a very rapid rate and blazing away with high luminosities," says Yale's Richard Larson, who has modeled the process. "The dark matter could then be remnants from the universe's very first burst of star formation—a host of unseen dwarf stars, neutron stars, and especially black holes. These first-generation stars may be more important than people give them credit for."

If not, this leaves "Jupiters," balls of gas too small to initiate nuclear burning deep in their cores, as the lone baryonic possibility. But there's a catch: No one can imagine how nature could have molded so many near-stars in a halo without making a considerable number of regular stars along the way. Theorist Hegyi looked into the problem and concluded that it's highly unlikely that dark, Jupiter-sized objects are hoarding all of the missing mass. "Even if ninety percent of the halo material is composed of Jupiters and only ten percent becomes stars, those stars would be giving off an enormous amount of light that we should be detecting with current instrumentation," he says. "So, if the dark matter is in the form of Jupiters, we're stuck with asking ourselves how all that mass got converted into Jupiters alone. Right now, that looks like an impossible job, but maybe we haven't been clever enough."

Given this current inability to finger with assurance any well-known celestial objects as the missing mass, many members of the astronomical community are convinced that the lion's share of the universe's dark ectoplasm, which is at least ten times the amount of luminous matter now visible, is not composed of protons and neutrons, but is instead made up of a more exotic species of matter, perhaps subatomic particles called neutrinos or axions postulated by particle physicists. This is an idea that could give men and women a cosmic-sized inferiority complex. Just as they were adjusting quite nicely to the revelation that Earth was not the center of the universe, they suddenly have to hear that they might not be made out of the cosmos's predominant material.

Arguments against ordinary matter as the missing mass are also being formulated in another corner of academia. Theoretical physicists these days have turned to the Big Bang as a giant laboratory to test their latest theories on both the creation and behavior of elementary particles. The results have important implications for cosmology and seem to indicate that baryons alone could not possibly have disguised themselves as dark matter. The equations that they are developing to describe that superheated fireball suggest that dark matter comprises, not ten times as much mass as the luminous stars and galaxies (which is the amount supported by current astronomical observations), but up to a *hundred* times as much. Cosmologists already know that such a huge quantity of additional matter could not possibly be in the form of protons and neutrons, because the very presence of those baryons in the early moments of the universe's cataclysmic birth would have greatly altered the production of such elements as helium and deuterium to levels not measured in our universe today. MIT's Alan Guth arrived at the factor of one hundred in 1979 when he sat down one evening in a Menlo Park, California, home and concluded that at birth the universe underwent a spurt of superaccelerated expansion that drove space-time (and concomitantly the universe's density) to the very brink between open and closed (see chapter 10)—so close to the borderline, says Guth, "that we wouldn't be able to measure, in our lifetimes, on which side we fall." Surety would not come until after a trillion trillion trillion lifetimes. If Guth is right, the most significant question in all astronomy, that of the fate of the cosmos, must remain unanswered.

Enthusiasm for the idea that the missing mass is other than the stuff of our everyday life grew immensely in 1980 when American physicist Frederick Reines and a Soviet group of experimentalists headed by V. A. Lyubimov independently announced that a bizarre elementary particle known as the neutrino appeared to have a very tiny mass (around a twenty-thousandth the weight of an electron). Previously, neutrinos were considered bits of "nothing," massless motes that carried away some energy during radioactive decay processes. Wolfgang Pauli first postulated their existence in 1930 in order to save one of the bulwarks of physics—the law of conservation of energy. Up to that point, it looked as if energy were actually disappearing during the decay of a radioactive atom. Pauli came to the rescue; the Austrian-American physicist reasoned that energy wasn't vanishing into thin air, but rather being carried off by a ghostly particle that had no mass when at rest, possessed no electric charge, and flew from the radioactive atom at the speed of light. Pauli's musings were confirmed in 1956 when Reines and Clyde Cowan proved the existence of the elusive particle during tests at a South Carolina nuclear power plant (an experiment appropriately named Project Poltergeist).

These "little neutral ones" are physics's most indifferent particles; they can travel through light-years of lead without being stopped. Physicists calculate that hordes of them were created during the Big Bang, outnumbering baryons, the particles of ordinary matter, by a billion to one. Trillions are passing through you right now as you read this sentence. Hence, giving neutrinos even the tiniest mass has enormous cosmological consequences, a possibility that theorists considered even before the Reines and Soviet experiments. They speculated that neutrinos with a bit of mass (scientists call them massive neutrinos) would not only account for the dark matter, but if abundant enough, their total mass would provide enough of a gravitational tug to bring our expanding universe to a screeching halt uncountable eons from now. All that is needed is about 0.000000000000000000000000000005 gram of matter in each cubic centimeter—the equivalent mass of three hydrogen atoms in every steamer-trunk-sized chunk of space. That doesn't seem like much to ask of the universe, but the luminous stars and galaxies counted so far can contribute only a hundredth that amount.

Throughout the early 1980s, massive neutrinos were truly the darling of the missing-mass set. They answered so many puzzles and satisfied so many cosmological hopes. Yakov Zel'dovich and his Soviet colleagues found them especially attractive since they provided the perfect mechanism for his top-down theory of galactic formation—the idea that immense clouds of matter called pancakes, spanning millions of light-years, emerged first after our cosmic birth and then later fragmented to form galaxies. Baryons alone couldn't form such gigantic structures, but neutrinos could accomplish the job with ease.

Zel'dovich's scenario begins tens of thousands of years after the Big Bang, when the universe is still a sea of particles. This sea, however, is not perfectly smooth. Reverberations from the Big Bang are "ruffling its surface," so to speak. Moving at near-light speeds, massive neutrinos wipe out the small-scale ripples, leaving only the very largest undulations to corral the phantom particles into a multitude of clouds, each cloud about a hundred million light-years across. That turns out to be the size of today's superclusters—a correspondence that, for many, has to be more than coincidental. It is assumed that these giant clouds of neutrinos act as gravitational traps. Ordinary matter, which settles out of the cosmic soup after the neutrinos do, is drawn by gravity into the ready-made neutrino clumps, like water into a whirlpool.

Astronomers and physicists alike were captivated by the notion that the smallest objects in the universe—elementary particles—could be controlling and shaping cosmic structures the size of superclusters. The two ends of the universal scale were linking up. Fascinated by the possible effects of a neutrino-filled universe, researchers in the Soviet Union, Europe, and the United States went to their computers to ask, "Given that neutrinos have substance, how would the universe evolve and what would it look like today?" It's an approach, a tool actually, that is relatively new to astronomy. As technology improves a computer's ability to tackle complex numerical calculations, astronomers now have the opportunity to recreate nature in a box of silicon chips —to mimic a realm that remains forever out of their direct reach. They can finally give that nebula or neutron star an experimental kick, if only in a computer memory. Some predict that the astrophysical insights that emerge from supercomputing may rival the profound changes wrought by Galileo's first use of the telescope. Simulating the evolution of the early universe is one step in this direction. Intriguingly, the various groups that worked on this problem discovered that a cosmos filled with massive neutrinos resembled certain aspects of our own.

Two such inquisitive researchers are Joan Centrella of Drexel University in Pennsylvania and Adrian Melott of the University of Chicago. Throughout the early 1980s, Centrella and Melott spent months of their time simulating and graphically displaying the evolution of a neutrino-filled universe in three dimensions. Their calculations were so formidable that they had to use the CRAY-1 supercomputer at the Lawrence Livermore National Laboratory in California. In one second, the CRAY-1 can handle more adding, subtracting, multiplying, and dividing than one person, working 24 hours a day hunched over a calculator, can perform in a lifetime.

With this electronic behemoth, Centrella and Melott fashioned a universe evolving hundreds of thousands of years after the Big Bang, a time when the neutrinos started wielding their influence. During the simulation, nearly one million imaginary particles represented the weakly interacting neutrinos; to the computer's eyes, they were spread out as a featureless haze throughout a three-dimensional grid. "Then we turned on the calculation, set the force of gravity to work, and saw how these particles began to act," explains Centrella. As the "eons" ticked away on the computer clock, gravitational interactions began to squeeze the neutrinos together at different rates in the three directions. If one axis collapsed first, a pancake was produced. If two axes collapsed faster than the third, a cigarlike filament formed. What was left behind were spherically-shaped voids, regions depleted of matter. You could even envision this process as voids growing and colliding, squashing matter between them and setting off galaxy formation. The result was a cellular structure of interconnecting pancakes and strings, somewhat like the Swiss-cheese appearance of our own universe described in the previous chapter. "And there's a flowing process going on," points out Melott. "The flow of matter is from the pancakes into the filaments and then into the knots where filaments intersect."

For a while, the riddle of the missing mass appeared to be on the brink of a solution. Bestowing mass on neutrinos was a more-than-adequate prescription. It seemed that the evanescent wisps of matter couldn't help but fabricate immense structures such as voids and filaments, so hard to explain in any other way. Unfortunately, the solution began to unravel. By 1983, the ghostly particles were revealing some troublesome Achilles' heels. First of all, it's difficult in the simulations to fragment neutrino pancakes into individual galaxies early in the universe's life. But, in the real cosmos, astronomers see globular clusters, galaxies, and quasars that are practically the age of the universe. And, as of this writing, additional experiments undertaken to confirm the Reines and Soviet findings have failed to detect a neutrino mass to everyone's satisfaction—and without mass, neutrinos are fated to whiz through the universe unabated, oblivious to ordinary matter.

A lonely vigil atop Arizona's Mount Hopkins resulted in further problems for neutrinos. There, University of Arizona astronomer Marc Aaronson used the innovative Multiple Mirror Telescope (MMT) to look closely at several carbon stars in the Draco dwarf galaxy, one of seven such tiny neighbors of the Milky Way (part of an endeavor that led him to entitle one of his papers "Carbon Stars and the Seven Dwarfs"). From Earth, these stars are so faint that Aaronson would sit before the telescope's digital display and actually count the individual photons arriving at the MMT from each star. "We literally rooted for each one," he jokes. Aaronson needed as many photons as he could get to figure out how fast the giant red carbon stars were moving within the dwarf galaxy, and, as with Zwicky's clusters and Rubin's spiral galaxies, the answer was a surprising one: The stars were much more stirred up than expected, which suggested that a cloud of dark matter must be permeating the Draco dwarf to keep the stars in check. But energetic neutrinos just can't settle down into a space the size of a dwarf galaxy. It would be like trying to corral a swarm of bees; it's hard to stuff them in. This is why neutrinos have been tagged "hot" particles. Many astronomers interpret Aaronson's finding to mean that the dark matter, whatever it is, must envelop individual galaxies, not entire superclusters the way neutrinos would congregate.

Lastly, difficulties arose as soon as researchers conducting the computer simulations began to ask, "Where might the galaxies be forming within the sheets of neutrinos?" While collaborating at the University of California at Berkeley, Marc Davis, Carlos Frenk, and Simon White saw that when the neutrino clouds did crystallize, the distribution of galaxies was unusually tidy. "We imagined ourselves sitting in one corner of this simulated neutrino-filled universe and picking out particles as if we were surveying galaxies in our own cosmos. We don't see neutrinos, but we do see galaxies," explains White, who continues this work from the University of Arizona. "In this way, we discovered that a neutrino universe produced voids and filaments much too cleanly and efficiently." When galaxies did emerge in the computer runs, they ended up huddling in dense knotty clusters that didn't resemble anything in the real sky. "Given these problems," says White, "it seems best to look elsewhere for a resolution of the nature of the dark matter. At least, that is, until experiment forces us to accept the existence of a massive neutrino."

White and others did not have far to look. Particle physicists are providing a cornucopia of additional suspects in the search for the missing mass. As investigators into the submicroscopic world of elementary particles refine their most current theories about the workings of that lilliputian domain, many new hypothetical particles pop out of their equations. Some of their models predict that every particle known to exist has a partner. For example, the photon might have its photino, the W boson its Wino, the quark its squark, and the graviton—the alleged mediator of the gravitational force—its gravitino. Neutrinos broke the dam. Researchers recognized that it needn't be just neutrinos that had mass; a host of other particles, as yet undetected and as elusive as the neutrino, could also have been produced in the Big Bang. University of Chicago astrophysicist Michael Turner wryly calls them WIMPs (for Weakly Interacting Massive Particles). Speculations on their existence opened up a whole new avenue of possibilities. Considering that physicists have been stalking subatomic particles for less than a century, only since the detection of the electron in 1897, it would be quite presumptuous to think that they've cornered them all.

One of the more popular WIMPs in physics these days is the axion. Physicist Frank Wilczek conceived of this wispy bit of matter to deal with certain aspects of the strong nuclear force (the force that keeps atomic nuclei from flying apart), and in a fit of whimsy not unusual for those in his esoteric field he named it after a laundry detergent! Axions make electrons look decidedly elephantine; they could be as much as a trillion times lighter. And if they do exist, a billion of them would be quietly poised in every cubic inch of space around us. Of course, "inos" and axions exist only on paper as yet, but there are plenty of other missing-mass contestants waiting in the wings, such as a varied assortment of fossil remnants that might have been left over from the Big Bang. They include quark "nuggets" (boulder-sized lumps of pure quarks, each nugget weighing as much as a planet), monopoles (weighty particles of magnetic charge), and tiny primordial black holes that may have formed during the first chaotic microsecond of the primordial explosion. Despite their seeming diversity, all the hypothetical specks and chunks of matter mentioned above have one thing in common: They wouldn't be as energetic, or hot, as neutrinos. Indeed, astronomers and physicists have come to call them "cold" dark matter.

At first it was thought that detecting cold-matter WIMPs directly, since they are so weakly interacting, would be nearly impossible. But physicists have been ingenious in devising various schemes to try to catch them. Axions, for example, should turn into countable photons of electromagnetic energy when they pass through an extremely strong magnetic field (a field almost 200,000 times stronger than the Earth's magnetic field). Other WIMPs might be detected as they enter a specially designed box containing trillions of superconducting micron-sized grains of tin or aluminum embedded in a nonconducting material. In principle, the WIMP would be sighted when it occasionally hit one of the grains and deposited some energy. The trick will be to distinguish that one collision from the myriad others occurring due to ordinary elementary particles passing through the box at the same time. Until such clever apparatuses are built, however, direct detection of any of these exotic candidates is just a possibility. Yet, if some form of cold dark matter does exist, it has already wielded considerable control over the evolution and structure of galaxies and clusters. This cosmological signature could be the strongest proof of its existence. Thus, the tables have been turned. Astronomers have long used the laws of physics to predict such things as the rates of fusion within a star. Now, physicists must turn to the heavens to look for clues as to what kinds of particles might be wandering through the universe undetected. Theorist Heinz Pagels calls it "the era of postaccelerator physics for which the entire history of the universe becomes the proving ground for fundamental physics."

One of cold dark matter's idiosyncrasies is quite tantalizing. Since these particles are supposedly not as frisky as a herd of bustling neutrinos, they can gather into smaller clumps after condensing out of the Big Bang—interestingly enough, clumps just about the size of a typical galaxy. The texture of the universe would then be woven in an entirely different manner than with massive neutrinos. Structure would grow from the small to the big: Peebles's bottom-up process at work. Sandra Faber of Lick Observatory, headquartered at the University of California's Santa Cruz campus, is one of many astronomers who have become enamored of this idea. A noted authority on the structure and evolution of galaxies, Faber never used to worry about anything smaller than a globular cluster, but subatomic particles have captured her fancy.

"A universe filled with cold dark matter turns out to have surprisingly precise predictive power for astronomy," says Faber. Her image of the early universe, a picture she helped fashion in collaboration with her Santa Cruz colleagues George Blumenthal and Joel Primack, is quite vivid: At first, the heavens look like a soup of baryons and cold matter, a very clumpy cosmic stew whose two diverse ingredients begin to separate. "When each clump hits the magic temperature of ten thousand degrees," explains Faber, "the baryons start to radiate away their energy, and when they do this, they sink to the center of the clump." The dark particles, which don't radiate, remain suspended in and around the baryons. What results is a core of ordinary luminous matter surrounded by a spherical halo of cold, dark particles—an object that looks suspiciously like a galaxy. And there's an added bonus to this model: Plain old thermodynamics limits the size of the galaxies that can be formed. Each clump would have to weigh between 100 million and 1 trillion solar masses for the baryon sinkage to be triggered. This covers all the galaxy sizes that astronomers observe today, from tiny dwarf galaxies to giant ellipticals. Later, when these galaxies cluster, their halos are expected to merge, the luminous disks and ellipsoids becoming embedded in the dark matter like white raisins in a chocolate pudding.

Cold dark matter even offers a plausible solution to that long-standing cosmological debate: When exactly did the universe decide to concoct two different types of galaxies—ellipticals and spirals—and why? "We believe that galaxies got differentiated from the very start," says Faber. In the collective mind of the Santa Cruz team, nature has the decided edge over nurture in the ultimate construction of a galaxy, with cold dark matter controlling the manufacture.

To support this view, Blumenthal, Faber, and Primack ask us to recall that picture of the early universe, right after the Big Bang, in which density fluctuations rush through the smooth ocean of particles. They imagine that some of these waves grow particularly high, sort of cosmic tsunami. "These turn out to be the regions where the cold-matter 'soup' is densest," says Blumenthal. "There is also not a lot of spinning motion." So, these are places where staid ellipticals quickly coagulate, like foamy whitecaps forming on a wavecrest. Ellipticals end up as slow-rotating bulges of stars simply because the enhanced density and lack of angular momentum in their neighborhoods inhibit the formation of any flattened, spinning disks of gas, not unlike our inability to twirl freely around in a packed, rush-hour subway car.

Spiral galaxies, on the other hand, tend to be loners, forming in less dense regions. This could be the reason that a spiral is a spiral. With room to breathe and lots of spinning motion in the gas, there would be no obstacle to the construction of large, gracefully swirling pinwheels of stars. It even allows for material to continue falling onto the disk through time so that spiraling arms can be maintained even to the present day.

Development of structure over larger scales is also possible in this model. The Berkeley team, among others, saw this during its simulation of an ever-expanding cosmos dominated by cold dark matter. It was by no means an easy task. Watching the universe develop over billions of years required up to seventy hours of computer time (by comparison, figuring a home-mortgage rate requires no more than a thousandth of a second of a computer's attention). At first, this miniuniverse, confined within the frame of a computer, appeared as a uniform mist of particles, "but then it became mottled," says White. "Very small clumps formed all over the place." Yet, at the same time, the cold-matter lumps started to line up into long structures. Those Big Bang perturbations coursing through the sea of dark matter, waves both large and small, were evidently working in concert with one another. This means that, with cold matter ruling the universe, a galaxy that's destined to end up in a supercluster may get an inkling of it in an early moment of the universe's history. The picture of the present-day universe generated in the Berkeley simulation, a network of voids and supercluster chains, closely resembles the real sky as depicted in the One Million Galaxies poster. But it is not quite the same.

The competition between the varied dark-matter candidates is far from over. As with neutrinos, cold dark matter has problems it too must overcome. One of the most nagging is its inability to carve out those gargantuan voids and spherical shells of galaxies sighted in the universe. Adding stars, dust, and gas to the "recipe," though, could change the outcome appreciably, since simulating an evolving universe, "hot" or "cold," without ordinary matter may well be like trying to bake a loaf of bread without the yeast. That one ingredient, stars radiating and dying, could make all the difference. Fully adding ordinary matter to the equations, however, requires stupendous increases in computer power. "It would be like trying to compute the path of every car driving around New York City, without neglecting the effects of each taxi driver's breakfast," quips White.

More problematic is a growing suspicion that the universe astronomers have been poking and probing these many centuries could be a delusion. All along, astronomers have been assuming that the visible galaxies serve as cosmic streetlights, tracing the way in which matter, both luminous and dark, is generally distributed throughout the heavens. But maybe not every dark-matter clump generates a luminous filling.

Visible galaxies could be like snowcaps in a mountain range, forming on only the highest (or, in this case, densest) "peaks" of dark matter. "If all your eye can see is the snow," points out White, "you end up missing all the other little hills." There could be lots of primordial stuff out there lying dormant, not being tallied up—very dim galaxies, for instance, that don't appreciably "light up."

Since the missing mass is a novel problem, a novel technique may be required to wrest additional clues out of the universe, and there's one on hand. Astronomers are an opportunistic lot, who often take newly discovered objects and fashion them into useful tools. Following this tradition, the prodigious list of missing-mass candidates has a chance at being pruned a bit by closely examining celestial mirages produced by a phenomenon known as gravitational lensing. Gravitational lenses were predicted to exist more than half a century ago, but were not actually detected until 1979.

The roots of this effect lie in Einstein's General Theory of Relativity, which said that a light beam would noticeably bend as it passed by a massive celestial body, such as the Sun. The Sun, in this case, becomes the gravitational equivalent of an optical lens. When astronomers, who were monitoring a 1919 solar eclipse, saw that starlight grazing the darkened Sun did indeed get deflected by essentially the amount predicted by general relativity, Einstein became world-famous overnight. Later, astronomers went on to speculate that nearby galaxies, with masses 100 billion times greater than our Sun, could produce an even more bizarre effect: Their gravitational fields would be strong enough to split the light of more distant objects into multiple images. Thus, galaxies would be nature's most powerful lenses.

Astronomers had to wait until 1979 to see their first gravitational lens at work, and only then through sheer happenstance. While routinely examining a photographic plate to locate the visible counterpart to a newly discovered radio source, British astronomer Dennis Walsh noticed that the radio object's position coincided with two starlike objects instead of just one. Additional observations confirmed that the cozy pair were not the chance alignment of a quasar and a star (as often happens), but rather the *same* quasar seen in duplicate. The light of this quasar, eight billion light-years distant, was being split in two by a giant elliptical galaxy located halfway between the quasar and Earth. You might think of the quasar light as a stream of water that comes upon a rock and gets diverted to either side of the stone.

"Many thought at first, 'Well, that's a nice trick, but it's not very fundamental.' But there's a dawning realization among many astronomers that gravitational lenses can do some very important work," notes Princeton astronomer and gravitational lens hunter Edwin Turner. "For one, they have the potential to distinguish between various dark matter candidates."

The idea is to carefully monitor the multiple quasar images to see if they "twinkle." Stars twinkle at night because their light is passing through turbulent patches of air that constantly shift the image about. Similarly, as the light from a quasar skims past a lensing galaxy, it too might flicker. If the dark matter in a galactic halo is primarily composed of faint stars or Jupiters, the quasar's image is expected to dim and brighten as the orbiting stars and planetoids pass in front of the quasar's light beam. Verifying this effect will require some patience; just one twinkling could take several years. Conversely, the lack of a shimmer could boost the idea that the dark matter is composed of submicroscopic elementary particles, objects too tiny and too smoothly distributed to affect the lensed quasar light at all.

"Although none of the gravitational lens systems discovered so far has led to significant progress on this problem, this fact should not be too discouraging," says Turner. "It might be recalled that the first three pulsars discovered did not teach us anything about neutron stars, stellar evolution, or general relativity, but larger samples eventually did. It may well be the same for gravitational lenses." As of this writing, less than a dozen gravitational lens systems have been uncovered, but ongoing radio and optical searches show promise of increasing the population appreciably. Their worth as a cosmological tool will be considerable once astronomers have a few dozen with which to work. They even have the potential to become cosmic-sized "optical benches" that enable astronomers to measure directly both the Hubble constant and the amount of deceleration, if any, in the universe's expansion.

When will this missing-mass mystery finally be solved? A bizarre cosmic particle shooting in from the far reaches of space might provide the answer tomorrow. But most astronomers believe it will take years, perhaps even decades, to sort through this cosmic jigsaw puzzle. "Astronomers are in the business of collecting clues," says Peebles, "but we're only in chapter two of this particular mystery, the point where the protagonist is sifting through the evidence to see which clues are relevant. We're a long way from the closing chapter."

For now, computer simulations can serve only as a guide, not a final proof. Despite the ardent desire of many cosmologists to have so much dark, exotic matter around that the universe will eventually stop expanding and collapse, current astronomical observations still say otherwise. Measurements of both the motions of galaxies within clusters and the spins of spiral disks support the same story: that though there is up to ten times more dark stuff than luminous matter it is nonetheless only a tenth the amount needed to halt the universe's expansion. Under those circumstances, "Jupiters" alone can still fill the bill as the missing mass.

Yet, if space telescopes peering out to the very edge of our visible universe see that cluster-sized clouds of gas, *not* galaxies, came into being first, then astronomers might be forced to conclude that massive neutrinos, or something like them, really do control the fabric of space-time. For that matter, the next generation of particle accelerators coming on-line, instruments whose miles-long tunnels will offer physicists the opportunity to explore energy realms never before attained, could reveal that one of those hypothetical "inos" is more than a mathematical squiggle on a piece of paper and perhaps bountiful enough to seal the universe's fate.

Of course, as Santa Cruz physicists Primack and Blumenthal once pointed out, the dark matter could be NOTA (that is, none of the above). Already, physicists on the threshold of unifying the four forces of nature under one set of laws are finding that their latest (and quite exotic) mathematical schemes hint at the existence of particles never before imagined. One of the most curious is something called "shadow matter." This odd material conceivably originated at the moment that gravity first broke away from its sister forces to evolve independently. Shadow matter would be a sort of doppelgänger to the ordinary matter we see and touch around us. There could be shadow electrons, shadow protons, and even shadow photons invisibly flying by and through us. If enough of this ghostly substance gathered nearby, we'd feel its gravitational pull but never ever see it. We wouldn't hear it, taste it, touch it, or smell it either. Gravity would be the only force that ordinary matter and shadow matter have in common. Our most sensitive electronic instrumentation would fail to detect its presence, since our world wouldn't share the same electromagnetic or atomic forces with the shadow world. Science-fiction writers could have a heyday imagining shadow planets revolving around shadow stars which orbit shadow galaxies eternally out of our reach . . . a phantom universe that inhabits our space-time. This missing mass would remain forever missing.

Or maybe the missing mass's secret formula has many ingredients. Already, several theorists are considering the possibility that heavy neutrinos formed clouds that stretch over entire clusters, forging the large voids and superclusters in the universe, while Jupiters and/or cold "inos" serve as the dark matter around individual galaxies. But few are eager to embrace this scheme, until it's absolutely necessary. According to "Turner's Law," the invocation of the tooth fairy should not occur more than once in any scientific argument. For economy's sake, astronomers are guided by that philosophic rule known as Occam's razor: When several theories are in competition, choose the simplest explanation first.

"In the meantime," says Vera Rubin, "my colleagues and I will have to learn to carry out our research with some amusement, recognizing that we are directly observing only five or ten percent of what the universe really has to offer."

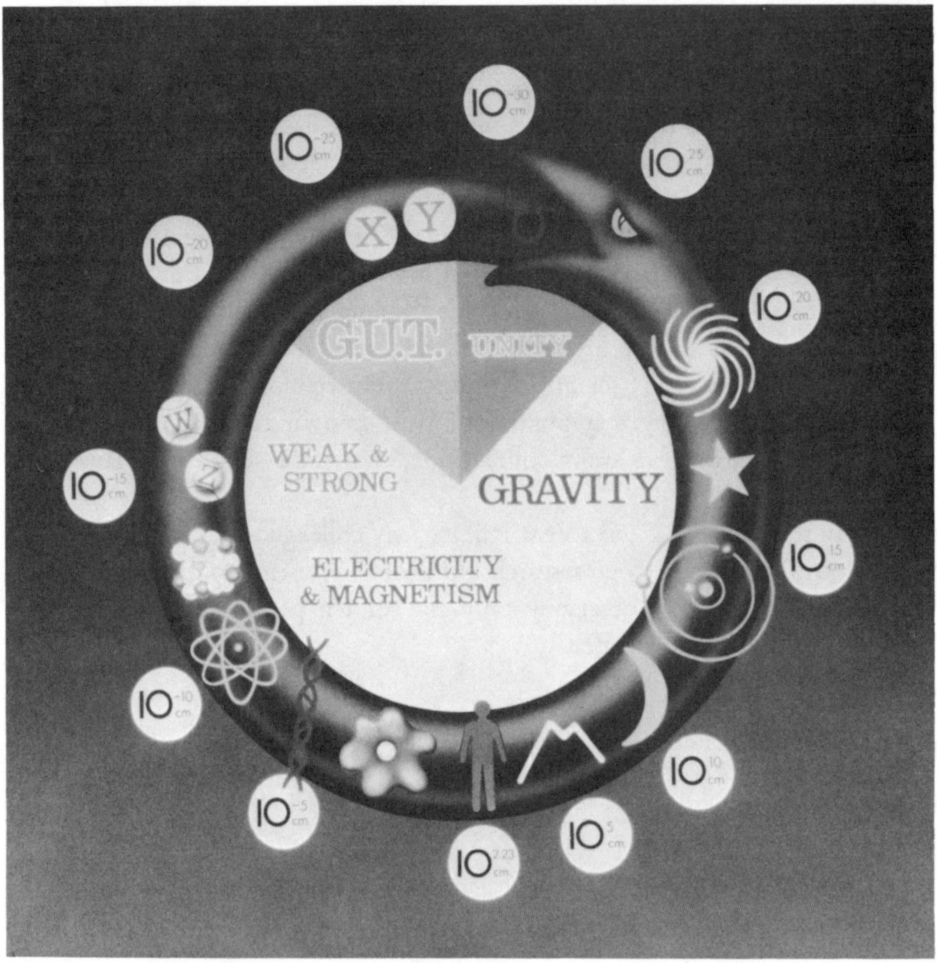

This diagram, using the ancient symbol of a snake eating its tail, was originally drawn by Nobel laureate Sheldon Glashow to represent the union of the macrocosm and the microcosm. It shows the ranges in distance—from cosmic to nuclear—over which the fundamental forces in nature prevail. *Robert Burger © Discover Magazine 6/83, Time, Inc.*

9

IN THE BEGINNING

I want to know how God created this world. I want to know his thoughts. The rest are details.

—Albert Einstein

"All roads lead to Rome," it was proudly proclaimed during the tempestuous reign of the Caesars. Likewise in astronomy, all inquiries and discussions inevitably lead to cosmology: the branch of astronomy that strives to explain the origin, evolution, and overall structure of the universe. Cosmologists are anxious to know how the universe began and how it concocted the vast amounts of matter that eventually condensed into the billions and billions of galaxies dispensed throughout a continually expanding volume of space-time. The entire cosmos is their laboratory; the birth or death of one star is but a minor detail.

In a guest lecture at Oxford, British astrophysicist Martin Rees commented that "cosmology is a peculiar science. It is by definition the study of a unique object and a unique event. No physicist would happily base his theories on a single and unrepeatable experiment, yet that, in effect, is what cosmologists are asked to do. No biologist would formu late general ideas of animal behavior after observing just one rat, which might have its own peculiarities. Yet cosmologists are not even, as it were, shown one rat doing the same thing over and over again—they are shown just one rat, doing just one thing, and asked to explain the behavior of rats on the basis of that observation." So extraordinary is this endeavor that some scientists still consider cosmology somewhat of a folly, especially now that cosmologists, by venturing further and further back through time in their theoretical reveries, are fearlessly attempting to reconstruct, microsecond by microsecond, the earliest moments of our cosmic genesis.

By the 1980s, while making these grand speculative leaps into the unknown, cosmologists came to realize that the key to cosmology, the science of the very large, actually lies in the science of the very small: particle physics. Theoretical physicists are excitedly advancing toward their goal of fashioning a single axiom that will unify all the varied particles and forces of nature, and, as a consequence, cosmologists are being brought ever closer to the very origin of the universe. It's not a one-way street. Particle physicists, many of whom once dismissed cosmology as an inexact science, are gradually discovering that constraints set down by astronomical observations are helping them decide whether new particles hypothesized from their equations can even exist in the observed universe.

Many researchers are convinced that the bubbly large-scale structure of the observable universe—roughly 10^{87} bits of light and matter scattered over 10^{69} cubic miles of space—can serve as a shadowy guide to the way in which nature originally forged its elementary particles: a clue to the amounts, the masses, and the types of atomic building blocks that emerged during the first instants of the universe's existence as a primordial fireball, whose smoldering ashes are viewed by radio telescopes.

Why, for example, should the universe be composed largely of matter, which current observations seem to support, without having equal quantities of antimatter hanging around? In a universe whose basic physical laws seem to stress balance and symmetry, one might expect similar portions of matter and its annihilative mate, antimatter, to have been forged in the Big Bang. But, then again, if the universe had acted with such equanimity, identical amounts of particles and antiparticles thoroughly destroying one another, we would never have evolved to make the observation. Cosmologists long scratched their heads over the preponderance-of-matter puzzle, but newly developed theories in particle physics are at last providing some enticing explanations for this life-giving bias in the universe's makeup.

Nobel laureate in physics Sheldon Glashow often likes to draw the ancient symbol of a snake eating its tail to represent this marriage of the macrocosm to the microcosm. The union, as we shall see, is proving quite fruitful as cosmologists proceed to figure out how the universe got here and where it is going.

Cosmology, as a strictly scientific pursuit, is a twentieth-century invention. Before the 1900s, thoughts relating to the universe's creation were largely the concern of theologians and philosophers. The ancient Babylonians, Hebrews, and Norse all referred in their early writings to a deity or deities that separated light from dark, land from sky, at the dawn of time. A fifth-century B.C. Greek philosopher named Anaxagoras declared that the cosmos must have arisen out of a primordial chaos—"an infinite aggregate of infinitesimal particles." Before the arrival of the twentieth century, evidently, an understanding of creation was limited only by one's imagination.

It was Hubble's discovery that the universe was ballooning outward that moved the search for the origin of the universe from the realm of metaphysics to that of science. With the systematic study of galaxies and their motions, cosmological models could finally be tested observationally. Actually, a few years before Hubble even linked a galaxy's redshift with distance and drastically altered our perception of the cosmos, certain theorists had already anticipated that the universe might be dynamic instead of static. In the early 1920s, for example, a young Russian meteorologist and mathematician named Alexander Friedmann boldly dropped the infamous (and short-lived) cosmological constant from Einstein's equations of general relativity and ended up with one solution in which the universe forcefully expanded, a result that Einstein had earlier rejected. With these results, Friedmann foresaw that galaxies should be moving away from the Milky Way with speeds proportional to distance. But, given the pre-Hubble prejudice for a tranquil, unchanging cosmos, no one took much notice. Soon afterward, Friedmann died of typhus while still in his thirties.

In 1927, unaware of Friedmann's work, Belgian priest and astronomer Abbé Georges Lemaître independently reached the same conclusion as the Russian mathematician and published the results in a little-known Belgian journal. This article, too, might have languished in obscurity, if not for the fact that the world's premier astrophysicist at the time, Arthur Eddington, eventually got wind of the idea and avidly discussed it with his peers, including Hubble.

To Lemaître, it was perfectly logical to assume that the expansion described in his calculations started somewhere. All he had to do was imagine the condition of the cosmos at earlier and earlier moments: By mentally putting the expansion of the universe into reverse, the galaxies move ever closer to one another, until they ultimately merge and form a compact fireball of dazzling brilliance. With this picture in mind, Lemaître imaginatively suggested that the universe began with the explosion of a "primeval atom," a state of extremely hot and highly compressed matter. Today's galaxies were just fragments blasted outward from this original sphere of material, which the Belgian cleric figured was about the size of our solar system. "The evolution of the world can be compared to a display of fireworks that has just ended: some few red wisps, ashes, and smoke," Lemaître would later write. "Standing on a cooled cinder, we see the slow fading of suns, and we try to recall the vanished brilliance of the origin of the worlds." From this poetic kernel of an idea would emerge the present-day concept of the "Big Bang," a model of creation that shapes the thoughts of astronomers today as strongly as Ptolemy's cyclic vision of the heavens influenced astronomers in the Middle Ages.

By the late 1940s, George Gamow, a Russian physicist who had emigrated to the United States, became the astrophysical community's most enthusiastic advocate of the Big Bang model and the scientist most closely identified with it. How the universe came to construct its varied assortment of atoms, from helium to uranium, was still a problem that vexed astronomers in the forties, but Gamow was sure that Lemaître's "cosmic egg" provided an answer. To Gamow's mind, the fiery miasma could have served as a marvelous soup in which the ninety-odd chemical elements were brewed in one fell swoop. Gamow liked to refer to the original stew of protons and neutrons as "ylem," after Aristotle's name for the basic substance out of which all matter was supposedly derived. In a whimsical takeoff on the biblical Genesis, Gamow once explained how deuterium, a form of hydrogen that has one neutron as well as one proton in its nucleus, originated: "In the beginning God created radiation and ylem. And ylem was without shape or number, and the nucleons were rushing madly over the face of the deep. And God said: 'Let there be mass two.' And there was mass two. And God saw deuterium, and it was good . . ."

Gamow's idea that "nucleosynthesis" (the creation of atomic nuclei) happened within some kind of primeval explosion was very useful indeed—at least for making the simplest elements such as deuterium. But Gamow's scheme, worked out in collaboration with his doctoral student Ralph Alpher,* turned out to be ineffectual in creating elements beyond helium. According to their calculations, the cosmic expansion generated by the "bang" both dispersed and cooled down the hot primordial ylem before heavier elements could form. "God was very much disappointed, and wanted first to contract the Universe again, and to start all over from the beginning," Gamow wrote in his lighthearted "Genesis." "But it would be much too simple. Thus, being almighty, God decided to correct His mistake in a most impossible way. And God said: 'Let there be Hoyle.' And there was Hoyle. And God looked at Hoyle . . . and told him to make heavy elements in any way he pleased. And Hoyle decided to make heavy elements in stars, and to spread them around by supernova explosions."

There were additional problems that put the Big Bang model under a cloud in 1948. Until Baade and Sandage readjusted the cosmic distance scale, making it clear that the visible universe is a much bigger and older place than previously estimated (see chapter 5), it was initially thought that the fateful cosmic explosion had to have occurred less than two billion years ago. "Clearly," conceded Gamow at the time, "a universe 1.8 billion years old could not contain five-billion-year-old rocks." This

*Their famous paper on this topic, though, listed the authors as Alpher, Bethe, and Gamow—unbeknown to Hans Bethe. Gamow couldn't resist the pun on the first three letters of the Greek alphabet: alpha, beta, and gamma.

and other difficulties spurred a group of Cambridge astrophysicists—Thomas Gold, Herman Bondi, and Fred Hoyle—to advance an alternate theory to the Big Bang: the "steady state" model of the universe.

The steady state universe was an expanding one (Hubble's data couldn't be denied), but it was a cosmos with no beginning and no end. It always looked the same, because matter was continually and spontaneously being created to fill in the gaps opened up by cosmic expansion. Galaxies were incessantly forming to replace those that receded beyond our view. The Cambridge group estimated that, to do this, only one atom of hydrogen needed to be created each year in every 35 million or so cubic feet of space, a volume roughly the size of New York City's Empire State Building. For steady-state supporters, creating a tiny bit of something out of nothing each year was scientifically much easier to swallow than imagining a monstrous primeval fireball originating all at once. "It seemed absurd to have all the matter created as if by magic . . . without a blush of embarrassment," said Hoyle. The steady-state model provided a safe haven for those who were squeamish at contemplating an honest-to-goodness origin for the universe. There simply was no beginning. A steady state universe always was and would always be.

Throughout the 1950s and beyond, a lively, sometimes bitter, discussion went on between the "big-bangers" and "steady-staters." Ironically, it was at this time that Hoyle, the avowed steady-stater, first described the explosive version of creation as a big bang during a British radio program on cosmology. The name stuck. More importantly, the debate served as a tremendous stimulus to astrophysical research. Prodded by their desire to avoid a primal cosmic egg, steady-staters figured out that stellar interiors and supernovae were marvelous environments in which to construct the weightier elements, as Gamow had conceded. Finding out that heavy elements are synthesized inside stars was a boon to astronomy no matter what the cosmology. This success alone, though, could not keep the steady state theory alive. As the dispute wore on, astronomical evidence continued to mount in favor of a Big Bang model of creation. The discovery of very bright quasars in the far reaches of the universe, for instance, strongly suggested that galaxies were evolving from some primordial state. And no one had ever observed an "old" galaxy next to a "new" one, as would be required in an eternal creation; most galaxies looked to be about the same age, which more than hinted that they formed simultaneously.

By the mid-1960s, these and other developments encouraged a whole new generation of physicists to consider once again the kinds of nuclear reactions that could take place during the first few minutes of a blazing genesis. Following well-known laws of nuclear physics, experimentally tested in laboratories such as William Fowler's at Caltech, they worked out the detailed mechanisms by which nuclear particles could fuse in a Big Bang, manufacturing the lighter elements.

The scenario that they wove from their theoretical and computer-calculated figurings began a few seconds after the primordial explosion, a time when the protons, neutrons, and electrons in existence were dashing helter-skelter through a very dense ocean of neutrinos and photons—"a thick, viscous fluid of light," as one theorist described it. The blinding flash of a hydrogen bomb would look like a dim night-light in comparison to this radiant maelstrom.

A significant change in the fireball, however, occurred at the 100-second mark. As soon as the expanding universe had cooled to about a billion degrees (nearly a hundred times hotter than the core of our Sun), the neutrons and protons could start sticking together and so build up a pool of deuterium. "The very instant that happened, the dam broke, and other nuclear reactions swiftly took place," explains Stanford University astrophysicist Robert Wagoner, who was a key member of that second generation which picked up where Gamow and company left off. Deuterium, a fragile thing, immediately combined with available protons and neutrons to form helium, which is very stable. A very small fraction of this helium, in turn, joined with other intermediate products of nucleosynthesis to forge lithium. The universe, during this brief spell, was no less than a giant fusion reactor.

"Within minutes, it was over," says Wagoner. "The chemical abundances were firmly set." One-quarter of all the mass in the universe was converted into helium; three-quarters remained as hydrogen nuclei (single protons); a fractional smattering of lithium, helium-3 (a rare form with two protons and one neutron) and leftover deuterium accounted for the rest. The stuff of people and planets, such as carbon, oxygen, and iron, came long afterward. With the universe continuously expanding and its density consequently plummeting, more complex nuclei could not be constructed until the birth of stars, millions and millions of years later.

Such a theoretical exercise appears to be a very brazen undertaking:

daring to describe the conditions of a minutes-old universe some fifteen billion years after the fact. But it turned out to be a very valuable exercise. Astronomers proceeded to measure the relative amounts of deuterium, helium, and lithium currently scattered throughout the universe. In laboratories, they analyzed the chemical composition of meteorites, pristine leftovers from the solar system's birth. With telescopes, observers searched for signs of primordial elements in the spectra of stars and of the gases drifting through interstellar space. Even an aluminum-foil "window shade," pulled down for a while on the Moon's surface by *Apollo 11* astronauts in 1969 and then brought back to Earth, enabled researchers to determine abundances from the solar-wind particles that collected on the shade.

So far, each and every measurement of a key light element has turned out to be remarkably close to the chemical abundance predicted from Big Bang nucleosynthesis calculations. "The amazing thing," points out University of Chicago astrophysicist David Schramm, who helped prove that deuterium can be made *only* in a Big Bang, not in stars, "is that the calculated lithium abundance is less than one-billionth of the universe's mass, while helium is twenty-five percent. We have predictions that range over ten orders of magnitude, and yet these amounts all agree with observation." Such an accurate match between theory and measurement has lent great strength to the Big Bang theory. At the same time, the observations convinced astronomers that the Big Bang could not possibly have forged enough ordinary matter, the stuff composed of protons and neutrons, to allow the universe to collapse in the far, far future. At most, ordinary matter provides only a few tenths the density necessary to stop the universe's expansion. That is why cosmologists who, for either philosophic or theoretical reasons, yearn for a closed universe must seek their solution in more exotic elementary particles, material that doesn't get involved in nucleosynthesis.

This work on nucleosynthesis, carried out throughout the 1960s and 1970s, was certainly a triumph for the Big Bang concept over its competitor. But the most devastating blow to the steady-state model of the universe was actually struck in 1965. The memorable showdown rested on an explicit and testable difference between the two opposing theories of creation. The Big Bang theory clearly predicted something that the steady state theory did not: a reverberation from the primeval blast that would echo throughout the length and breadth of the universe to this day.

Gamow was the first to suggest that the bountiful flood of highly energetic photons released in the aftermath of a Big Bang process would cool down with the expansion of the universe and appear today, in every direction of the celestial sky, as a uniform wash of radiation. Many astronomers like to think of this faint glow as the cooling embers from that cosmic fire of yesteryear. In 1948 Gamow's colleagues, Robert Herman and Alpher, estimated that the overall temperature of the waning fire would by now have dropped to within several degrees above absolute zero. The lingering warmth would be detectable as centimeters-long microwaves. No one, though, did anything about this fascinating prediction, despite its usefulness as a tool for determining which of the cosmologies was correct. Radio astronomy, which could verify the prediction, was still in its infancy at the time, and cosmological tests were not high on its list of priorities. And with the Big Bang model then on the outs, Gamow never pushed to make the observation; so, the prediction fell into obscurity. Most astronomers forgot it altogether.

The idea resurfaced in the 1960s, when Robert Dicke at Princeton University, as well as Zel'dovich and his colleagues in the Soviet Union, again reasoned that residual heat from the Big Bang must still be permeating the universe. Upon coming to this conclusion, Dicke encouraged two colleagues to construct the equipment that would enable them to measure this pervasive hum of radiation. Meanwhile, P. James E. Peebles, also at Princeton, calculated the expected temperature of the heat. But in the process of setting up their antenna on a campus rooftop in 1965, the Princetonians learned that two radio astronomers with Bell Telephone Laboratories, Arno Penzias and Robert Wilson, had already been listening to the faint primeval noise—and quite unwittingly. The discovery of the universe's microwave background is one of the most oft-told tales of serendipity in astronomy. But as the famed French chemist Louis Pasteur long ago noted, "In fields of observation, chance favors only the mind that is prepared."

In 1964 Penzias and Wilson had begun to calibrate a massive horn-shaped antenna located in Holmdel, New Jersey, not far from the site where Jansky first detected celestial radio waves. The three-story-high antenna, which looks just like a gigantic measuring scoop, was originally built to receive intercontinental radio signals bounced off Echo balloon satellites. It was Penzias and Wilson's job to convert the communications receiver into a telescope that could measure extremely faint radio waves radiating from beyond the plane of the Milky Way.

During their initial tests, though, the antenna's liquid-helium-cooled receiver consistently registered excess noise, no matter where the instrument was pointed. Converting the energy of that excess microwave radiation into an equivalent temperature, it appeared as if the entire sky were weakly heated to nearly three degrees above absolute zero. Over a year's time, Penzias and Wilson ruled out innumerable candidates for the spurious three-degree signal, from atmospheric radiation to man-made interference emanating from nearby New York City. At one point, the tenacious astronomers, who refused to dismiss the problem as simply system noise, even considered a biological explanation. "There were a couple of pigeons living in the antenna that had deposited the usual pigeon droppings," points out Wilson. "But only a minor improvement occurred when we cleaned up the droppings and disposed of the pigeons."

Despairing that he and Wilson would ever locate the origin of the noise, Penzias chanced to mention the problem to a friend, who happened to have heard of Dicke's prediction via the scientific grapevine. As the cliché goes, the rest is history. Penzias soon invited the Princeton group to visit the Holmdel installation, just a few dozen miles from the university, whereupon it was confirmed that Penzias and Wilson had indeed been listening to the faint, remnant echo of the Big Bang. The ever so slight yet maddening discrepancy was the fossil whisper of creation. "To be honest, neither of us took the cosmology very seriously at first," recalls Wilson. "I had come from Caltech where Fred Hoyle's influence was very strong. Philosophically, I actually preferred the steady state cosmology. But, we were just happy to get some kind of explanation for that troublesome noise." They were more convinced of the discovery's cosmological significance once they saw a write-up of the finding on the front page of *The New York Times.* Any remaining doubts certainly disappeared by 1978 when the two Bell Lab astronomers received the Nobel prize for their serendipitous discovery, one of the few times the physics prize has been awarded for an astronomical observation. Others had actually detected the cosmic noise earlier, but had written off the extra heat as a systematic error in their instruments. Little did they realize that they had been three degrees away from science's most coveted award.

Most cosmologists agree that Penzias and Wilson's accomplishment was the most important cosmological advance since the discovery of the universe's expansion by Hubble. As soon as the cosmic microwave background was confirmed, the Big Bang theory became the undisputed winner in the battle of cosmologies. Indeed, confirmation of the three-degree radiation was what spurred many to reconsider the problem of Big Bang nucleosynthesis and to calculate its products in more detail. From that point on, support for the steady-state viewpoint rapidly dwindled, until mention of it essentially disappeared from the scientific literature.

The microwaves now bathing the universe take astronomers back to a specific stage of the universe's creation. For a few hundred thousand years after the cosmic explosion, with its initial burst of nucleosynthesis, the expanding fireball was a murky potpourri of protons, simple nuclei, electrons, and most of all neutrinos and photons. If astronomers could somehow peer back to this epoch, they wouldn't see much, for the cosmic plasma was actually opaque. No sooner was a photon emitted or (more likely) scattered by a particle than it was absorbed or scattered by another bit of matter. Consequently, the universe at this time was a blurry soup impossible to see through, just as the Sun's hot outer layers prevent us from gazing at its core. This is why no matter how sophisticated the instrumentation, optical telescopes could never see back to the Big Bang itself. The so-called "fireball" would be an impenetrable barrier to our view.

This opacity was drastically altered, though, once the universe was about half a million years old. At that stage, the cosmos had attained a temperature low enough (approximately 6,000 degrees Fahrenheit) for the negatively charged electrons to start combining with the positively charged protons and nuclei. Drawn to the nuclei by electromagnetic attraction, the electrons were finally able to surround the nuclear particles without getting kicked out of orbit by an energetic photon. Thus, what was once a jumble of dissociated particles became a more settled collection of neutral atoms.

Because atoms of hydrogen and helium interact very little with light, the photons now found themselves able to travel through the universe unimpeded. As a result, the primordial fog lifted, and the universe became transparent for the first time. The light waves, mostly in the optical and infrared region of the spectrum at this point, were gradually stretched with the universe's expansion, until today they are detected as that vast sea of microwaves, regularly dispersed throughout space. It is the amazing uniformity of the microwaves (their distribution over the sky varies by less than one part in 10,000) which informs astronomers that the primeval fireball was exceedingly smooth, as well as dense and hot.

Over millions and millions of years the backdrop of the universe, illuminated by the relic photons, changed from yellow, to red, and finally to the deepest black as temperatures dropped. All the while, the universe with which we are most familiar was taking form and shape. Within a billion years, the atoms somehow clumped into clouds (the exact mechanism, as we saw in chapter 7, is still the subject of lively debate), and the clouds eventually organized themselves into galaxies. The most energetic newborn galaxies, the quasars, announced their debut by spewing brilliant fountains of energy from their central cores.

The cosmic microwaves falling on terrestrial radio telescopes disclose the conditions of the universe a half-million years after the Big Bang, and the nucleosynthesis calculations indirectly push back the frontier to within one second of the explosion. But what was the universe like a millionth of a second after its birth? A trillionth of a second? And at what point did the protons and neutrons come into existence?

"Nucleosynthesis permitted cosmologists to look back to the first second, but speculating about the earlier evolution of the universe was a no-man's-land," recalls Gary Steigman, who has carried out many pioneering calculations on the conditions of the early universe at the Bartol Research Foundation, located on the campus of the University of Delaware. "In the 1960s, we didn't have the physics to describe the behavior of the universe at earlier times. For example, it had been conjectured that there was a maximum temperature to the fireball, about a trillion degrees. But there were philosophical problems with that conclusion: Why should the universe have started out, ready-made, at that temperature?"

The lack of an adequate model describing the universe's condition prior to one second also kept another cosmic mystery from being solved. Standard theories of physics in the 1960s suggested that nature, which seems to adore parity, would have created equal amounts of matter *and* antimatter during the Big Bang. But, after many decades of looking around the universe with an assortment of equipment, astronomers have detected only matter-galaxies, matter-stars, and matter-planets (at least, so far). What happened to the bulk of the antimatter?

Possible answers to many of these cosmological riddles at last arrived as theoretical physicists closed in on a long-sought goal: the unification of the four basic forces that control the actions of everything from galaxies and asteroids to raindrops and silicon chips. On the macroscopic level, there are the familiar forces of gravity and electromagnetism; on the microscopic level, the less-familiar strong and weak nuclear forces. Each force acts very differently at the low temperatures of our normal surroundings, but, say theoreticians, they all become identical when energies are high enough. Because these energies, though, are far beyond the reach of any foreseeable particle accelerator, particle physicists have had to turn (grudgingly at first, but then more eagerly) to the field of cosmology. They are now convinced that the forces of nature were assuredly one during the first hellish moments of the universe's creation. The Big Bang has become to them the ultimate particle accelerator—a cosmic-sized test lab for their newest and most challenging theories.

Cosmologists in turn have come to understand that by observing how particle physicists piece together their theoretical jigsaw puzzle, they will be offered a tantalizing peek at the cosmic soup that existed an infinitesimal fraction of a second after the Big Bang. To best comprehend this fascinating new marriage between cosmology and particle physics, though, we must first identify the plethora of particles that make up the subatomic world and the forces that reign over them.

Although science novices gazing at the arcane formulas in a physics textbook may have their doubts, the job of physicists can be summed up very easily: to describe the workings of the universe and all its complex and various phenomena in the simplest way possible. When

Newton realized that the set of equations describing the trajectory of a rock thrown into the air could equally be applied to plotting the orbit of a planet or comet around the Sun, complexity gave way to simplicity. He had discovered that the motions of these two disparate objects, a small earthbound rock and an entire planet, were controlled by one, all-encompassing force—gravity. Gravity is the most far-reaching of the forces, capable of wielding its tugs over trillions of miles, yet it is also the feeblest interaction. As physicist James Trefil notes, "An ordinary magnet can lift a nail, even though the entire Earth is on the other side exerting a gravitational attraction."

A magnet exhibits one of the three interactions added to the roster of forces since Newton's day. Electromagnetism, the most conspicuous force on the human scale whether displayed in a lightning bolt or a dime-store compass, keeps electrons buzzing around a compact knot of protons and neutrons, thereby enabling atoms to exist. Without atoms, there would be no chemical elements, and without chemical elements there would be no life. And by discerning an even deeper structure within the nuclei of atoms during the first few decades of this century, physicists came to recognize the "strong" and "weak" forces. The strong force keeps every atomic nucleus, a packed assembly of protons and neutrons, from flying apart in the face of electromagnetic repulsion, while the weak force controls the way in which certain subatomic particles will disintegrate, causing a nucleus to radioactively decay. The weak force is also involved in nuclear fusion, so vital in powering the stars. (Could there be other forces? Possibly. Physicists would not be too surprised if new, very short-range forces, cousins to the strong and weak nuclear forces, are uncovered as particle accelerators increase their energies.)

Just as the number of forces acknowledged by physicists grew from one to two to four, so too did the so-called "elementary" particles proliferate. In the 1930s, physicists confidently thought of electrons, protons, and neutrons as the sole units of matter. But as atom smashers grew over the intervening decades from crude room-size mechanisms to instruments with miles-long tunnels, myriads of ephemeral particles appeared in the debris spewed forth as protons and electrons crashed into stationary targets at near-light speeds. For a while, it was hard to keep track of all the new species. Such variety was bewildering to the scientists who were seeking simple truths, not intricate conundrums, from their energetic probes into the heart of the atomic nucleus.

A theoretical breakthrough in organizing the vast particle array was made in 1963 when Murray Gell-Mann and George Zweig independently suggested that many of the puzzling particles might actually be composites, each a different combination of smaller, more fundamental constituents. Gell-Mann called these basic building blocks *quarks,* alluding to a line from James Joyce's *Finnegans Wake:* "Three quarks for Muster Mark." In the lexicon of physicists, so noted for their quirky labels, these quark types, now numbering six, are identified as *up, down, charm, strange, top,* and *bottom* (although more poetically minded physicists prefer to call the last two *truth* and *beauty*). Using this scheme, the proton turns out to be composed of two "up" quarks and one "down" quark. The neutron, the proton's near-twin except for a lack of electric charge, is a composite of two "downs" and an "up." While no one has indisputably observed a naked isolated quark, a number of accelerator experiments have strongly hinted that protons, neutrons, and a host of short-lived nuclear particles are indeed made out of pointlike entities.

Aside from quarks, the only other structureless and indivisible bits of matter presently thought to exist are the electron, the electron's close relatives the muon and tau particles, and three kinds of neutrinos. Collectively, these six related particles are known as *leptons,* from a Greek word meaning small. Nature contrives diverse constructs from its fundamental constituents. Particles of matter, it turns out, are either leptons or quark composites, and this simplification was a tremendous boost to certain cosmological inquiries. "The quark model finally enabled physicists dealing with the Big Bang to go further back in time, beyond the protons and neutrons," explains Steigman. "That funny maximum temperature, once thought to exist, was really just a signal that a phase transition was occurring at that point—a transition from protons and neutrons to the quarks themselves." Additional strides toward understanding this uncharted territory of the Big Bang, the first microsecond of creation, were made as theorists conducted a parallel search for an underlying order among the four forces of nature.

Theoretical physicists have come to believe that the four interactions —gravity, electromagnetism, and the strong and weak nuclear forces— are just different manifestations of one ancestral force. It's analogous to the way that diamond, charcoal, and graphite are different expressions of a single substance. A sparkling gem, a blackish clump, and a greasy lubricant certainly look and feel different to us, but at some level they are essentially the same: All of them are carbon. Likewise, it is theorized that the four forces once exhibited a basic similarity within the fiery kiln of the early universe.

This process of theoretically unifying the forces actually began in the 1860s when James Clerk Maxwell, a Scottish physicist known as Daffy to his school chums, consolidated electricity (the force that governs the output of a lightbulb) and magnetism (the force that gently swings a compass needle northward). Maxwell, by putting the experiments of Michael Faraday in mathematical form, saw that the two forces were merely different sides of the same coin, each unable to exist in isolation. Electric currents are always accompanied by a magnetic field, and conversely a variable magnetic field generates electricity. Maxwell's equations even predicted that electromagnetic energy could travel through space as an undulating wave, a realization that by the twentieth century led to radio, television, lasers, and microwave ovens.

Einstein had high hopes of continuing this process of unification by joining electromagnetism with gravity in one mathematical construct. He devoted a good part of his life to this quixotic pursuit, but, alas, to no avail. In some ways, he jumped the gun, for two other fundamental forces, the strong and the weak, first needed to be discovered and fathomed.

But where Einstein stumbled, a new generation of physicists were able to push forward. By the late 1960s, Harvard's Glashow, Steven Weinberg, now with the University of Texas at Austin, and Abdus Salam of the Imperial College of Science and Technology in London all made (future Nobel-prize-winning) contributions to showing that there was a fundamental and intimate link between electromagnetism and the weak nuclear force. We certainly don't perceive this "electroweak" interaction in our relatively frigid environs; the unification of electromagnetism with the weak force occurs only at very high energies. Within the framework of the Big Bang, the electroweak force exerted its influence directly on the cosmos when the primeval fireball was no more than a ten-billionth of a second old. After that infinitesimally short stroke of time, electromagnetism and the weak force took on their separate guises: the weak force limited its reach to the dimensions of an atom, while electromagnetism developed a much longer range.

Many years had to go by, however, before the Weinberg-Salam-Glashow model could be fully accepted (initially, in fact, the theory was virtually ignored). Particle accelerators first had to become powerful enough to directly reach the high-energy domain—around a hundred billion electron-volts* or an equivalent temperature of 1,000 trillion degrees—wherein the electromagnetic and weak forces join up. Definitive proof came during the closing months of 1982 when a pencil-thin beam of protons, racing clockwise within a four-mile-long underground accelerator ring at CERN, the European Center for Nuclear Research near Geneva, slammed head-on into a focused beam of antiprotons traveling in the opposite direction. Whenever a proton and antiproton happened to crash into one another and completely annihilate each other, a monstrous state-of-the-art detector, weighing some 2,000 tons and as big as a house, recorded the resultant shower of debris—the newborn energy readily coalescing into a fresh batch of particles. More than one hundred physicists from thirteen laboratories around the world were involved in the endeavor. By 1983 the head of the CERN team, Harvard's Carlo Rubbia, was able to announce that a couple of handfuls of their observed collisions, which numbered in the tens of millions, generated the distinctive signatures of the W^+, W^-, and Z particles, the predicted "carriers" of the electroweak force. "There's now such a sense of confidence. It doesn't seem as if we are making it up as we go along," said theorist Weinberg at the time of the discovery.

That an atomic particle can be conceived as the conveyor of a force was one of the revolutionary outcomes of modern physics. When walking about on the Earth's surface, we often get the impression that a force is some kind of invisible entity that pushes or pulls us around. But on the level of atoms, physicists prefer to describe forces as a kind of tennis game: A force between two particles arises from their continually exchanging another, identifiable particle (a subatomic tennis ball, so to speak). For electromagnetic interactions, the "tennis ball" is the

*Particle physicists traditionally describe accelerator energies in terms of electron-volts. One electron-volt is the energy an electron picks up as it crosses a one-volt electric field, the voltage in a penlight battery.

photon. The Z and W particles, meanwhile, are responsible for transmitting the weak force. And, as an expression of the strong force, something called a gluon constantly bounces between quarks to bind them into protons and neutrons. In keeping with this stratagem, a particle called the graviton, not yet detected, is thought to convey the force of gravity.

Such a division of labor means that there are actually two basic categories of elementary particles in the universe: the particles that give rise to the forces and the particles that serve as the building blocks of matter. The force particles are known as *bosons,* while the building blocks have come to be called *fermions.* Each category obeys a different set of physical laws, too. Bosons can bunch up and merge, in accordance with Bose-Einstein statistics (hence, the name). Fermions, on the other hand, are relegated to certain definable slots within the nucleus or atom, as if they were assigned to specific desks in an elementary-school classroom; they are forbidden to pile up (a type of ordering described by Fermi-Dirac statistics). That is why two beams of light, comprised of squeezable bosonic photons, can pass right through one another, but fermionic people cannot. The quarks and electrons in us just won't budge from their allotted positions.

The success of the electroweak theory gave physicists, Howard Georgi and Glashow being among the first, the courage to push their unification schemes to energies as high as a trillion trillion electron-volts, a realm where the electroweak force at last merges with the strong force. Altogether, the various mathematical models that attempt to describe this unification are referred to as Grand Unified Theories, or GUTs for short (although some think it a pretentious name, since gravity is not included). Leon Lederman, director of the Fermi National Laboratory, a facility which operates the United States' most powerful atom smasher, the Tevatron, says that their ultimate goal is "to explain the entire universe in a single, simple formula that you can wear on your T-shirt"—a sort of superadvanced version of $E = mc^2$. When the weak and strong nuclear interactions are united with electromagnetism in a single force, it is believed that quarks and leptons, such as electrons and neutrinos, become virtually indistinguishable, quickly and easily changing from one form to the other. It is a capability which, at one crucial moment in the universe's history, may have resulted in some cosmos-shaking consequences.

Unfortunately, unlike the electroweak merger, the GUT unification occurs at such high energies that it is technologically impossible to duplicate the effect on Earth. "Energies as high as those involved in GUTs are far beyond what we can get in terrestrial laboratories," points out Schramm. "It would require an accelerator that stretches from here to Alpha Centauri, which would ease vacuum leak problems but would make data analysis difficult—not to mention problems with the gross national product."

But luckily, physicists have access to this high-energy arena through a convenient back door. "There was at least one event that took place at these energies: the Big Bang itself," continues Schramm. "The first 10^{-35} second of the universe's history provides our best testing ground for the grand unifying ideas." While discussing the growing importance of cosmology to particle physics during a climb in the Dolomite mountains of northern Italy one summer, Schramm (the astrophysicist) and Lederman (the particle physicist) decided to set up a special cosmology group right at the Fermilab site, just outside Chicago. The added presence of cosmologists at a particle accelerator laboratory, a move replicated at other facilities around the world, serves as a powerful symbol of the new partnership that has been forged between the once diverse fields.

"In the early seventies, you could have counted on the fingers of one hand the number of people applying cosmology to particle physics," adds Steigman. "But then, particle physicists realized that they were going to have to use cosmology to constrain all the free parameters in their models. The first suggested model for the axion particle, for instance, had to be ruled out, because it would have affected the structure of red giant stars in a way that is not observed. Nowadays, when physicists write down a new model and look at the particle content, they are asking, 'Is it consistent with Big Bang nucleosynthesis? Is it consistent with the density of the universe?' Not too long ago, for doing that, I was often referred to as a 'half-astrophysicist.' But now, there's no turning back. Workshops on the early universe are so dominated by particle physicists that astronomers can barely sneak in."

A visitor to such meetings will often hear these new cosmologists describe the embryonic universe as swiftly proceeding through a series of "phase transitions," each stage altering the early universe's basic physical properties (analogous to the way that water is physically transformed as it cools from vapor to liquid to solid ice). During the first searing flash of creation, it is believed that the four forces of nature were united. But then, as the universe coursed outward and cooled, the individual forces (and their associated collections of particles) broke away one by one, each force eventually assuming its own identity.

Under the rules of this theoretical game, physicists came to see that the force of gravity was the first to part company, 10^{-43} second after the initial explosion. Though vastly important on cosmological scales, it ultimately became the weakest of the forces; the gravitational force between an electron and proton is 10,000 trillion trillion trillion times feebler than the electrical force binding those same two particles in an atom. By the time the universe was about 10^{-35} second old, the GUT unification was shattered, allowing the strong force to develop its own characteristics. A fleeting instant later, the electroweak force divided into its two separate components. All the while, the various particles out of which we are composed congealed, like crystals of ice in a cooling pond of water. Verifying this fascinating tale of creation can be tricky, because the Big Bang laboratory shut down more than 10 billion years ago. Unable to construct a particle accelerator from here to Alpha Centauri, Big Bang specialists must be content to search for supporting evidence in the many fossils that remain behind from that most ancient of epochs.

Early on, cosmologists recognized that the material universe itself is actually the most viable and visible relic of the Big Bang process. We and everything around us are by all appearances composed of matter. The cosmos seems to have a decided aversion to keeping large quantities of antimatter in tow, and for many years the reasons for this preference were not readily apparent. The very existence of antimatter was first hypothesized in the late 1920s by the prolific British theorist Paul A. M. Dirac. On the occasion of his eightieth birthday, Dirac recalled that "the whole beauty of the mathematics [of uniting relativity with quantum theory] would have been spoiled" without postulating that every type of elementary particle had a mirror image, a sort of photographic negative that was equal in mass but exactly the reverse in such properties as electric charge and spin. And should any particle ever meet up with its complementary antiparticle, the two would completely annihilate one another in a burst of pure energy. "We are agreed that your idea is crazy," a colleague told Dirac upon hearing of the startling conjecture. "What divides us is whether it is crazy enough to be true." It was. By 1932 the telltale track of an antielectron, or "positron," the electron's positively charged counterpart, was sighted in a cloud chamber during a cosmic-ray experiment. Since then, the list of bona fide antiparticles has grown quite lengthy.

If the universe were initially so "symmetric" (as physicists like to describe the original unified state), one would think that the Big Bang would not have been partial and would have made just as many antiparticles as particles. The universe, in this way, would be half matter and half antimatter. Indeed, collisions within particle accelerators always seem to yield equal amounts of matter and antimatter. But Steigman, who was one of the first researchers to press this issue, points out that astronomical observations consistently argue for an all-matter cosmos. Outside of an atom smasher, antimatter is almost as rare as a penguin on a tropical isle. The fact that our space probes have never been obliterated upon landing on the Moon or the planets is pretty good evidence that the solar system is safely constructed out of matter. If the Viking lander had set down on an antimatter Mars, for example, the blinding explosion would have dominated the nighttime sky. "And the gamma rays being emitted within our galaxy and from nearby clusters of galaxies set limits on how much matter is currently mixing with antimatter and annihilating," points out Steigman. "On the scale of clusters of galaxies, millions of light-years, the results tell us that only one particle in a million could be antimatter. And the limit is more than a million times lower for our own galaxy. This has convinced many people that the universe is not at all symmetric."

There are, though, dissenting opinions. "Are there cosmologically significant amounts of antimatter, even whole galaxies of antimatter, elsewhere in the universe?" asks NASA physicist Floyd Stecker. "The truly conservative answer to this question at this point is that we do not know." Using GUT equations, Stecker has constructed a model of the Big Bang in which separate matter and antimatter domains are created, producing a universe that resembles a honeycomb, some of the "cells" harboring matter exclusively, the others having predominantly antimatter. Stecker's suspicion would be confirmed if higher-resolution gamma-ray telescopes detected streams of energetic photons being emitted along lengthy seams in space, a signal that matter and antimatter were specifically destroying one another along the far-off boundaries of the separated regions. Others would be convinced if an entire antinucleus (the two antiprotons and two antineutrons of an antihelium nucleus, for instance) plummeted to Earth as a cosmic ray. That would imply that antielements are indeed forming in antimatter stars situated in some distant antimatter galaxy. But, until such proof arrives, most physicists continue to believe that the universe overwhelmingly favors matter and point to the fact that grand unified theories, for the first time in the history of cosmology, offer them a plausible explanation for this cosmic prejudice.

Some of the universe's laws and subatomic inhabitants already display a clear-cut bias: Time proceeds in only one direction; radioactive cobalt emits many more left-spinning electrons than right-handed ones; and exotic particles called K mesons seem to fancy one mode of decay more than another when they disintegrate. Taking these observed preferences into account, Soviet physicist Andrei Sakharov went on to surmise that the early universe itself became a bit skewed, evolving in such a way that the cosmos ended up with more particles than antiparticles, even though the universe started out with equal proportions of both. But in 1967, when he first proposed this strange idea, it didn't fit any established theory; so, the suggestion was completely forgotten. Yet once grand unified theories came onto the scene, not only was Sakharov's proposal revived, but the matter/antimatter imbalance turned out to be a natural consequence in many of the models. Only in hindsight did physicists realize that Sakharov was onto something.

The era of grand unification allegedly occurred when temperatures in the newborn cosmos were still above a billion billion billion degrees. And, according to the theory, supermassive particles called X bosons, mediators of the combined weak-electromagnetic-strong force, were present in vast numbers. Each was roughly as heavy as a virus. Through the intervention of these X bosons (and their partners the anti-X bosons), quarks and antiquarks, leptons and antileptons were able to rapidly switch from one species to the other. Quarks could turn into antileptons or antiquarks; antiquarks into leptons or quarks.

But, like some game of musical chairs, the frenzied switching had to stop as soon as temperatures fell and the unification was broken. It was the moment when the universe chose sides. You could think of the fluctuating mixture of particles and antiparticles freezing into place as soon as the cosmic clock reached 10^{-35} second. As a result, for every 10 billion or so bits of matter and antimatter that settled out of the teeming cosmic soup, it is postulated that one extra particle of matter was made. Just like the K mesons, the X bosons and anti-X bosons exhibited a discernable preference in the way that they decayed. When the "GUT music stopped," the anti-X bosons produced roughly 10,000,000,001 specks of matter for every 10,000,000,000 bits of antimatter generated by the X bosons. Within a millionth of a second, the quarks and antiquarks paired off and annihilated one another, releasing a blizzard of radiant energy. But, thankfully, that tiny surplus of unpaired quarks remained to create the material universe that both surrounds us and comprises us. This also explains how there came to be about 1 to 10 billion photons in the universe for every particle of matter, a ratio that physicists could only take on faith before the advent of GUTs.

"All of us who worked on this problem discovered that it was much easier to produce a matter/antimatter asymmetry than people had previously imagined," comments Fermilab cosmologist Edward Kolb, who was a major contributor to the model through his computer simulations of particle interactions in the early universe. "Being familiar with the computer codes written for nucleosynthesis, the reaction rates for turning neutrons and protons into nuclei, we adapted the codes to deal with quarks. The statistical mechanics is very similar, and, as with nucleosynthesis, the expansion of the universe plays a very key role.

"You have to pull things out of equilibrium," he goes on to explain. "If the universe were not expanding, or were expanding very slowly, then it would always have equal numbers of particles and antiparticles. To go back to your musical chair analogy, it is the expansion which pulls the chairs away."

Immediately after the quark/antiquark annihilation, the unconsumed quarks combined into groups of three to form protons and neutrons. During the next second of expansion, the electrons and positrons staged their own conflagration, annihilating and leaving behind a counterbalancing residue of electrons. Forming first light nuclei, then atoms, all of these leftovers from creation would proceed in the ensuing eons to coalesce into the galaxies, stars, and planets.

If grand unified theories are correct in this description of creation, the mechanism that allowed an asymmetry to creep into the universe should still be at work and thus provide a means of proving the conjecture. If a quark can suddenly change into an antiquark or antilepton, it means that certain quark composites such as protons, long thought to be eternally stable, have the potential to transmute and decay at random, suddenly shooting out sprays of lighter particles. To Glashow it means that "diamonds are *not* forever"; to the universe-at-large, a somber affirmation of the mortality of matter.

When the universe was younger than 10^{-35} second, transformations between quarks and leptons supposedly occurred all the time, since X bosons, the presumed middlemen in the exchange, were plentiful in that torrid habitat. It is presently estimated from preliminary experiments that today, in our much colder environment, where X bosons are virtually extinct, the quarks in a proton might take more than 10^{31} years to convert. (From the laws of probability which govern atomic processes, the stated lifetime of a particle, such as the proton, is really an average; half of the proton population would actually disintegrate sooner, the other half later.)

A rough handle on the lifetime of the proton comes, perhaps surprisingly, from medical science. The fact that the typical human body is not habitually riddled with cancer caused by internal bombardment from disintegrating protons tells physicists that protons must have a mean lifetime greater than 10^{16} years. Any less than that and we would be walking sources of radioactivity.

To determine the lifetime of the proton more precisely, physicists have always had two choices: either they can pick out one proton and patiently check whether it decays over a period of time 1,000 billion billion times longer than the age of the universe, or they can assemble a vast array of protons, about 10^{34} or so, and determine whether a few out of that horde fall apart spontaneously within a year or so. Obviously, the latter is the only reasonable approach, and research groups have been conducting such searches around the globe: in a converted garage along the seven-mile-long Mont Blanc automobile tunnel that runs between Italy and France; in a zinc mine situated more than half a mile below Mount Ikenoyama in Japan; in a century-old gold mine in southern India; in a salt mine near Cleveland, Ohio, 2,000 feet beneath the shores of Lake Erie. Despite the varied locations, a common thread runs throughout all of these experiments. Each group has to place their collection of protons (either a room-sized stack of iron or a vast pool of purified water several stories deep) below ground away from disruptive cosmic rays and then surround the proton reservoir with intricate electronic equipment designed to spot the distinctive flash of radiation from a dying proton.

Catching a proton in the act of disintegration would be potent testimony that theorists are indeed on the right track in their attempts to intertwine the electromagnetic, weak, and strong forces. At the same time, physicists would be witnessing an event that has not been prevalent since the origin of matter. It seems ironic that one of cosmology's outstanding riddles, the dearth of antimatter in the universe, might be solved with an underground tank of water and/or thick pile of metal, far removed from any glimpse of the heavens that are the raison d'être of these scientists.

The notion that each and every proton in the universe will eventually crumble, albeit at a preposterously slow rate, is surely a disturbing thought as it concerns the long-term fate of the universe, but it is a relatively tame consequence when compared to the other bizarre effects imagined with the new cosmologies. Many of the theories that unify the forces find it conceivable that, during the universe's initial expansion and subsequent cooling, points, strands, and/or walls of highly concentrated energy (remnants of the early universe's former symmetric state) remained after the universe experienced its rapid series of phase transitions, weaving a network of cosmological "cracks" in the topology of space-time reminiscent of the patchy flaws that crop up as a body of water freezes.

The one-dimensional faults, known as "cosmic strings," are one of the more interesting defects hypothesized. One might think of them as extremely thin tubes of space-time, skinnier than an atomic particle, in which the energetic conditions of the primeval fireball still prevail. Created before the universe was one second old, any strings that survived to this day would be either exceptionally long (spanning the entire width of the universe) or bent back on themselves creating closed loops that continually lose mass-energy by vibrating at velocities approaching the speed of light.

Theory suggests that loops of cosmic string can come in a variety of lengths and masses, depending on the particular phase transition in which the string was created. The most astounding, though, would surely be those strings conjectured to have been created around 10^{-35} second after creation, the moment when the GUT symmetry was broken. Each and every inch of this energy-packed string would weigh more than the island of Great Britain. One complete loop, which would run at least a few hundred thousand light-years around, could outweigh a small cluster of galaxies.

If these particularly potent strings truly exist, astronomers may not be too desirous to observe them close-up. While such a string, anorectically thin, could actually whiz through your body without bumping into one atom, its peculiar gravitational field would wreak havoc nonetheless: If this string sliced through you, your head and feet would proceed to rush toward one another at 10,000 miles per hour. These properties are certainly dramatic enough to have testable consequences. "The tremendous tension in the string, for instance, would make it wiggle around like a rubber band, producing lots of gravity waves," explains cosmic-string expert Alexander Vilenkin. Such gravitational radiation emanating throughout the universe could very well affect the timing of radio pulsars. Massive strings would also be excellent candidates for gravitational lensing, perhaps even producing a line of multiple quasar images.

Other loops of string that are both lighter and shorter, having survived from a later phase transition, would not exhibit such strong gravitational effects but would still have a chance at being detected through their electromagnetic interactions. While passing through our galaxy, for instance, a small string might appear as a radio-emitting filament.

These kinks in space-time also provide an alternate explanation for how galaxies, concentrated islands of matter, might have emerged out of a smooth primeval plasma. Vilenkin has proposed that stringy loops of the proper size, scattered throughout the early universe, could have attracted vast assemblages of matter around themselves, serving as the seeds of future galaxies. And the middle of each clump of gas could even have become dense enough to form a black hole, an object already thought to inhabit many galactic cores. Slowly shrinking as it oscillates and releases gravitational energy, each galaxy-forming string would eventually disappear from space-time, leaving only a spiral or elliptical behind as its calling card.

Over the years cosmology has enjoyed great success with its concept of an explosive birth. At an international meeting of astronomers, Soviet theorist Zel'dovich declared, quite confidently, that the hot Big Bang model is "as well established as celestial mechanics."

But, even as late as the seventies, there were grounds for caution in that assessment, for certain features of the universe's makeup remained, to everyone's frustration, unexplained. As one example, cosmologists could only guess at the reasons for the universe being so amazingly smooth over very large scales. By some means—no one knew exactly how—the universe got filled with a nearly uniform and intensely hot gas in thermal equilibrium, which stayed relatively homogeneous as the universe evolved. Inexplicably, the cosmos seems to be as finely tuned as a precision tool-making machine. It was not until 1979 that a possible explanation for this curious initial condition arrived, and ironically the solution was introduced by a physicist with absolutely no background in astrophysics. To resolve some of cosmology's most perplexing mysteries, this budding researcher stumbled upon the idea that the universe did not merely expand at the moment of its birth, but that it also burped!

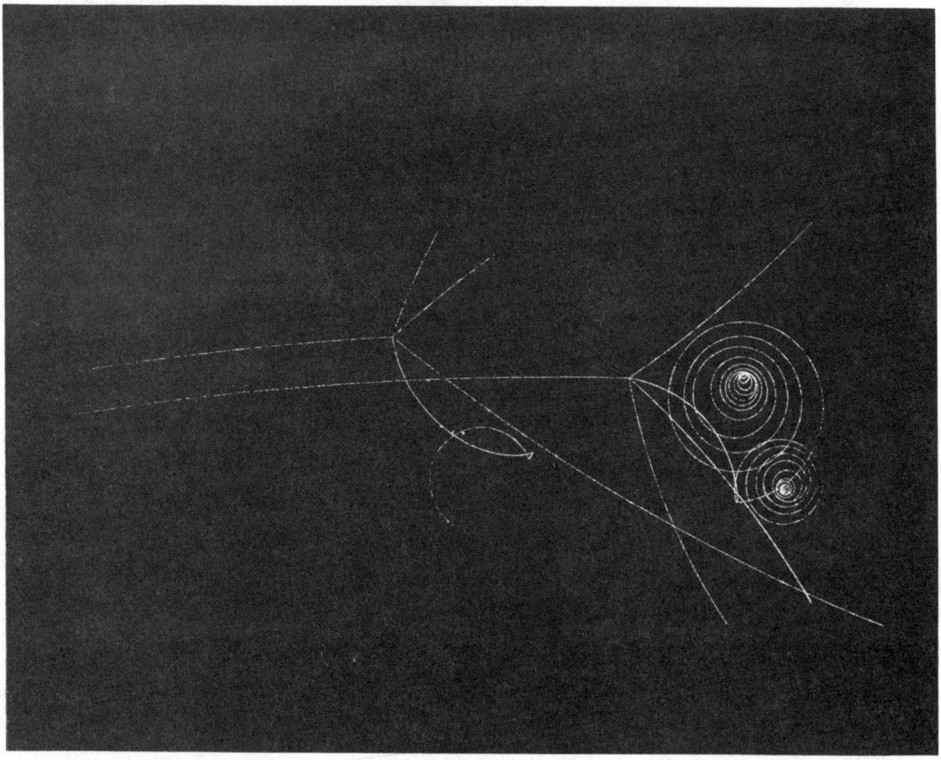

According to a revolutionary new theory, every particle in the cosmos is simply the residual glow from an early, momentary burst in the expansion of space-time. Within a bubble chamber, one of those particles, a proton, collides with a pion, creating an electron, which spirals clockwise, and its antimatter mate, a positron, which spirals counterclockwise. *Courtesy of Lawrence Berkeley Laboratory.*

10

THE COSMIC BURP

It is often said that you can't get something for nothing. But the universe may be the ultimate free lunch.

—Alan Guth

Alan Guth was a most unlikely candidate to alter our understanding of the universe in one night. He was one of physics's many gypsies, an intrepid "post-doc" who wandered from university to university filling temporary lectureships and research positions after obtaining a Ph.D. from the Massachusetts Institute of Technology in 1972. His first stop was Princeton, where he served as an instructor in particle physics for three years. From there, he went on to Columbia University and then Cornell, all in all an eight-year trek from the woodlands of central New Jersey to the Finger Lakes region of western New York.

During this time, Guth was oblivious to cosmology, almost as unacquainted with its basic tenets as a college freshman. "Frankly," he says with the bright smile that often crosses his face, "I thought it was too speculative." Instead, Guth was immersed in studying the many and varied forces of nature, not experimentally by the side of a giant particle accelerator but rather from a theorist's perspective. Guth was one of that legion of theoretical physicists who attempt to describe the forces that govern every action of the universe's many inhabitants, from the farthest galaxy to the smallest subatomic particle, in the most elegant mathematical language possible. His graduate thesis attempted to show how quarks, those basic units of matter, might be joining up to become protons and neutrons, the core ingredients of an atomic nucleus. He had no interest in hazy conjectures on the universe's emergence from a vantage point fifteen eons down the road. Mathematical rigor was his passion.

Yet, serendipity—three times over—would eventually nudge him toward the study of cosmology. Two lectures, a chance collaboration with a colleague, and an off-the-cuff remark would lead him to a late-night revelation that dramatically revised the whole "industry's" model of our cosmic birth. Guth's calculations reveal that our universe may have begun, not only with a bang, but with a sort of cosmic *burp* —a brief moment of superaccelerated expansion that transformed a subatomic smudge of energy into a celestial cornucopia of galaxies, stars, and planets.

If Guth's fundamental concept is correct, it also means that the universe astronomers have long studied through their telescopes is only a minuscule mote immersed in a much larger domain of space-time. In the sixteenth century, Copernicus displaced humanity from the center of the universe by suggesting that the Earth revolved around the Sun. Nearly four centuries later, Hubble continued the process by proving that our beloved Milky Way is but one in a myriad of galaxies rushing through the vast gulfs of outer space. Now Guth was moving us one more step into obscurity by suggesting that the cosmos visible to astronomy's mighty array of instrumentation is only a speck when measured against a larger stage of space-time.

Guth's ideas have had considerable impact on the astrophysics community because they appear to solve several cosmological mysteries that have plagued theorists for years and to answer some very basic questions about our explosive beginnings: Why the universe was once so hot, why it keeps expanding, and where it obtained its supply of mass and energy.

Guth's unplanned scientific odyssey began on the afternoon of November 13, 1978, with the young researcher sitting in an auditorium on the Cornell campus listening to a visiting lecturer, Princeton theorist Robert Dicke, expound on cosmological paradoxes. Why, Dicke asked his audience, raising one of cosmology's thorniest puzzles, is the universe so geometrically flat? There's another way of viewing this problem: Why is the universe virtually poised, like a ballerina *en pointe,* between open and closed?

It was Einstein who first taught cosmologists of the intimate relationship between matter, gravity, and the curvature of space-time: Matter, it's been proved many times, causes space to warp and bend. If there is not enough matter in the cosmos to exert the gravitational muscle needed to halt the expansion of space-time, then our universe will remain open, destined to expand for all eternity. With too little mass, space never turns inward, closing back up on itself. Instead, a mass-poor space curves out like a saddle whose edges go off to infinity, fated never to meet.

On the other hand, a higher density would provide enough gravity to lasso the speeding galaxies—slowing them down at first, then drawing them inward until space-time curls back up in that Big Crunch and reforms the brilliant fireball from which we were spawned. Here, space-time would encompass a finite volume and yet have no boundaries. The two-dimensional analogy would be the surface of a sphere like our Earth which, as sailors discovered hundreds of years ago, has no edges. If a space voyager traveled a straight-line course through a closed universe long enough, she'd eventually return to her starting point (not unlike a round-the-world cruise here on our own planet).

But astronomers cannot yet predict with absolute certainty which fate, eternal expansion or fiery collapse, will befall us; as we saw in the chapter on the missing mass, the density of matter now measured in our universe lies relatively close to the threshold, a point that cosmologists call the critical density. At this notable juncture, space-time looks neither completely open nor closed, but flat as a pancake. Working with the standard model of the Big Bang, Dicke, in collaboration with P. James

E. Peebles, realized that this was a bewildering situation. Since any deviation from critical density at the moment of creation would grow quite rapidly with time, Dicke explained that to be within ten percent of critical density today (as measured by observers) means that the density of the universe at the age of one second had to differ from critical density by less than one trillionth of one percent. At the time of Dicke's talk, there was simply no explanation for this incredible fine-tuning, as difficult a task as throwing a dart from New York City and hitting a bull's-eye on a dart board held over Washington, D.C. "I didn't understand the basis for Dicke's statement at the time," admits Guth. "But it did strike me as an amazing fact."

Around this same time, a friend and colleague at Cornell, Henry Tye, began asking Guth whether he thought grand unified theories would give rise to magnetic monopoles. "I had heard the words 'grand unified theories,' before, but I knew nothing about them, absolutely nothing," recalls Guth. "Henry explained to me what they were, and I went home one night to think about whether monopoles would be a natural outcome from this theory."

Guth was familiar with the concept of a monopole; he had studied it before on an abstract, mathematical level. As with antimatter, Paul Dirac first predicted the monopole's existence more than half a century ago when he contemplated nature's many symmetries. If the universe has provided us with separate units of electric charge—the positively charged proton and the negatively charged electron, for example—then it's likely, Dirac surmised, that it also cooked up separate particles of magnetic charge.

This is a concept that goes against the grain of our daily experience. Every magnet on Earth is comprised of an inseparable duo: a north and south pole. Break the bar in two and you only end up with two new magnets, each with its own north and south pole. This would be true even if you continued to break each piece down to its very last atom. But *mono*poles, as the name implies, would be elementary particles that have only one pole, either north or south. In 1982, the entire physics community was abuzz when a sophisticated detector set up by Blas Cabrera in a tiny basement lab at Stanford University appeared to see a monopole passing through its supercooled coil. As of this writing, the signal has not been repeated, so the jury is still conferring on whether this alleged magnetic debris from the Big Bang is truly coursing through the universe. Many hope it is. "Roses are red, violets are blue, the time has come for monopole two," a playful group of Harvard theoretical particle physicists once wrote Cabrera.

However, it was 1979, and Guth was not concerned with the experimental evidence but rather the latest theoretical grounds for justifying a monopole's existence. From the perspective of GUTs, monopoles were simply the pointlike analogs of cosmic strings: infinitesimally small spots in space where the conditions of grand unification continue to prevail. "Once I understood the rules, it didn't take too long to realize that grand unified theories would give rise to magnetic monopoles. In fact, monopoles with ridiculously high masses." To reach that conclusion, Guth used methods developed by Dutch physicist Gerard 't Hooft and Moscow physicist Alexander Polyakov, who a few years earlier had worked out a theory of monopoles with a simpler model. According to the calculations, each bit of magnetic charge would weigh more than a million billion protons. This was the mass of an amoeba squeezed into a volume that was smaller than a proton, making a magnetic monopole the superheavyweight champ of all the elementary atomic particles.

Tye was quite pleased by the finding and eager to figure out how many of these goliaths would have come spewing from the Big Bang. Guth, however, was more than a little reluctant. "In all honesty," he confesses, "I thought it was a sort of silly thing to work on." But again, a visiting lecturer, this time soon-to-be Nobel laureate Steven Weinberg, opened Guth's mind to the formidable possibilities to be found in combining particle-physics theory with cosmology. "Steve was one of the first to use grand unified theories to see how baryons [protons and neutrons] were produced in the Big Bang, and his talk left me with the distinct impression that the Big Bang was indeed a tractable mathematical problem."

Once on track, Tye and Guth saw rather quickly that the Big Bang would be quite prolific in its monopole production. In fact, a little too prolific.* "So many of these heavy monopoles should have been produced that we began to wonder why the universe was here at all,"

*A result independently reached and first published by Caltech's John Preskill, a Harvard graduate student at the time.

remembers Guth. "Their tremendous weight would have closed the universe back up eons ago." What happened to prevent the universe's early demise?

With the arrival of autumn Guth traveled west to fill yet another postdoctoral position at the Stanford Linear Accelerator Laboratory in California. But his collaboration with Tye would continue through phone calls and correspondence. They concluded that there was only one way the Big Bang could have avoided flooding the universe with heavy magnetic monopoles: It must have experienced a period of "supercooling," a sort of stall in the break-up of the grand unified force. In the mid-1970s Harvard theoretical physicist Sidney Coleman, among others, had already pointed out that this was a distinct possibility.

In the preceding chapter we saw how the familiar forces of nature, electromagnetism, and the strong and weak forces, were at one point united in the earliest moment of our cosmic birth. If you could somehow have done the impossible and sneaked a detector into this unusual realm, you'd have found that quarks, electrons, and neutrinos, distinct entities at the low temperatures of our everyday life, virtually looked and acted alike. Each force broke out of this cocoon of uniformity only as the universe expanded and cooled. Some like to think of this process as a sort of crystallization, as water crystallizes when it becomes ice. But maybe, posed Tye and Guth, the cosmos maintained its symmetry for a while as the temperature plunged, just as water can sometimes supercool and thus remain liquid below its freezing point. This would prevent too many monopoles, pointlike imperfections in the developing fabric of space-time, from arising in the "crystal" that was freezing out to become our universe.

Guth, sitting in an office whose neatly shelved books and organized papers reflect the order so often sought by the theoretical physicist, explains: "When you cool water it forms ice, and this ice forms along axes which are chosen at random. So, you have the possibility that one part of the ice will start forming with the crystal axis pointed a certain way, while another piece nearby points a different way. An imperfection in the crystal arises where these two pieces join. In particle physics, the creation of monopoles is the analogy of that process." Supercooling provided the time to smooth out these many misalignments and keep the number of monopoles to a minimum—at least few enough not to squish us down in a Big Crunch anytime soon.

Left at that, Tye and Guth's finding would have provided an interesting, though not momentous, footnote in the developing marriage of cosmology and particle physics. But Guth went one step further, spurred, he believes, by a casual reminder by Tye to check how this supercooling might have affected the expansion of the infant universe, the development of space-time itself. At the time, Guth little realized the import of that remark. In fact, he didn't get around to the job for quite a while. The spark may have been a long chat on grand unified theories that he had one Thursday afternoon with Harvard's Coleman, who was on sabbatical at the Stanford laboratory that year. Whatever the reason, the varied pieces of the cosmological puzzle that had been gathering in Guth's mind over the previous six months finally fell into place later that night, December 6, 1979.

Around eleven o'clock, Guth sat down in his makeshift office at his rented home in Menlo Park, a town near the Stanford campus, and began to work on a series of calculations that within a couple of hours would cover four pages. The title at the top of the first page records his ambitious intentions: The small, precise black letters announced that he was tackling no less than the EVOLUTION OF THE UNIVERSE. What follows is a potent blend of particle physics theory and general relativity, which has long served as the very backbone to all cosmological ponderings. "Actually, it's an easy calculation once you know the question to ask yourself," Guth modestly professes. "I'm surprised it wasn't done earlier."

In actuality, there were some foreshocks. Once a revolutionary idea has caught on, a search through the scientific literature will often point out precursors—the germ of the theory sprouting in earlier forms. In the 1970s, for instance, Soviet theorists Zel'dovich and A. D. Linde each wrote on the possibility of an early universe that wildly raced outward for a short time. However, they offered no special mechanism for this to occur. And at the very moment that Guth was formulating his model, others were deriving it independently. Theoretical astrophysicist Demosthenes Kazanas of the NASA Goddard Space Flight Center in Maryland would write in the *Astrophysical Journal* three months before Guth's paper was published in the *Physical Review* that "the presence of a phase transition early in the history of the universe, associated with spontaneous symmetry breaking . . . significantly modifies its dynamics and evolution. . . . The expansion law of the universe then differs substantially. . . ." This same point would also be made by a Japanese researcher named Katsuhiko Sato and University of Michigan physicist Martin Einhorn. But Guth's paper, with its greater detail, would serve as the definitive catalyst to this new line of research. "It's an interesting example of the sociology of science," points out Princeton's Peebles. "There are preliminary quakes, and then one paper catches your attention." Guth himself helped the process along by becoming a scientific barnstormer, widely lecturing on the idea before he even published to convince the doubting Thomases that his scheme was a powerful tonic for many of cosmology's ills. As Tye jokes, Guth had become a "born-again cosmologist."

Guth's four pages of equations look deceptively easy, yet any practicing theorist reading over the pages, now inconspicuously ensconced in one of many cherry-red binders lined up on Guth's bookshelf, would see immediately that Guth was dealing with the arcane tools of their trade—things called gauges, Higgs fields, and false vacuum states. Often it is difficult for physicists to explain, in everyday language, what those technical words mean. But, as Guth put down his pen around 1 A.M., the bottom line was undeniable. If his equations were valid, the universe did not just expand at the moment of its birth, it tore outward like a fanciful science-fiction spaceship in warp drive. Perhaps inspired by the double-digit rises in the cost of living at the time, Guth soon came up with an appropriate name for this period of hyperacceleration—he called it inflation.

At first glance, this may seem to be a subtle change, a mere alteration in the mechanics of the Big Bang. Why does it really matter that the universe once expanded at a much faster rate, you might ask? But it was much more than that. Guth's inflationary universe rewrote the script of our cosmic beginnings, giving us the first hint as to the origins of the matter and light that fill the universe.

The scenario begins 10^{-35} second into our birth, when what could be called Gamow's ylem was only one trillionth the size of a proton (about 10^{-25} centimeter across). This unimaginably hot seed was expanding and thus starting to cool below 10^{27} degrees. That's more than a million million million times hotter than our Sun, whose core temperature is practically absolute zero in comparison with such energies. Under the standard model of the Big Bang, this is the stage at which the grand unification should have started to fragment into separate, distinct forces and differentiated subatomic particles. But, as Guth and Tye suggested earlier with the monopoles, this did not happen right away.

Instead, the little knot of space-time became supercooled as the temperature plunged—again, just as water can sometimes remain liquid below 32° Fahrenheit before freezing into ice. This delay in its "crystallization" endowed the universe with a tremendous potential energy, not unlike a rock about to fall from a precarious perch on the edge of a precipice. What Guth realized late that night was that there would be peculiar side effects to this supercooling, most importantly on gravity. A pressure contribution to gravity, a term usually ignored in everyday computations, became very important in the early universe. In this bizarre supercooled state, the pressure term actually reversed the effect of gravity. In other words, gravity, normally a force that draws things together, did a turnabout and became repulsive, causing space to balloon outward at a superaccelerated rate. Within an infinitesimal fraction of a second (about 1/100,000,000,000,000,000,000,000,000,000,000 of a second) our observable cosmos doubled its size a hundred times over (2, 4, 8, 16, . . .) until it was the size of a softball, or maybe even larger.

This momentary inflationary epoch at last ended quite climactically when the supercooled symmetry (using physics's jargon) began to spontaneously break. At this point, the analogy to supercooled water can be taken one step further. When water freezes, it radiates a certain amount of energy into its environment. This is its latent heat and represents the difference in thermal energy between the solid and liquid states. A jostling drop of liquid water, it is readily apparent, contains more energy than an immobile piece of ice, and this extra energy is released during the transition. This energy is not minuscule. If the water in an Olympic-sized swimming pool were to freeze, from the surface to the bottom, it would give off enough energy to heat a large house for a couple of years. Theoretically, the energy released from a tray of ice cubes would be sufficient to run a refrigerator light bulb for a couple of hours. Unfortunately, this is not really possible. To turn that heat into a useful form of energy, say electricity, takes work.

But a similar energy difference does exist between the inflationary and noninflationary phases. Guth surmised that, upon freezing, the inflationary universe converted all its latent energy into an awesome cascade of extremely hot matter—in fact, *all* the particles and radiation that surround us today. It would be this fireball, not the earlier expanding ylem, whose glowing embers appear today throughout space as a cool wash of microwave radiation. Whatever mass-energy was contained in the preinflationary seed was simply overwhelmed by the fiery flood tide.

Surprisingly, Guth figures that only about twenty pounds of hot, symmetric mass-energy is needed to get the process going. Thus, it was inflation's demise that put the *bang* into the Big Bang and provided our cosmos with all its necessary building materials. According to this scheme, every galaxy, dust cloud, and photon—10^{87} particles in all—is only the resultant mass-energy from that brief but frenzied inflationary era. As Guth likes to put it, "Our universe is the ultimate free lunch." Mountains and asteroids, pulsars and nebulae, people and pine trees, every material entity in the universe is simply the residual afterglow of inflation.

After the release of inflation's pent-up energy, the superacceleration stopped, gravity went back to being an attractive force, and expansion of the heavens continued at a more sedate pace as space-time coasted outward on the sheer momentum left over from that initial, hyperexplosive thrust. With this theory physicists could understand, better than ever before, why the early universe was so hot and how it got so big.

Guth didn't comprehend these features all at once that balmy December night, but he did smell the air of truth about his initial calculations, quite an accomplishment for someone unacquainted with cosmology only the year before. The next day, after a record bicycle ride to work, the very first words the excited researcher wrote in his notebook were: "SPECTACULAR REALIZATION: This kind of supercooling can explain why the universe today is so incredibly flat—and therefore resolve the fine-tuning paradox pointed out by Bob Dicke in his . . . lectures." As if to back up his judgment, he drew a double box around the paragraph.

How inflation solves the flatness problem is quite intuitive. Just imagine the surface of a balloon as it's being blown up. As the balloon gets larger and larger and larger, its curvature gets flatter and flatter and flatter. And a universe doubling its size every 10^{-35} second only magnifies this effect in a preposterous manner. By the end of the inflationary period, the universe's curvature was greatly suppressed, taking it to the very brink between open and closed. Since the geometry and density of the universe are intimately linked, Guth's findings were also telling astronomers that there should be a *hundred* times more matter wandering through the cosmos than currently viewed through telescopes. As we saw in discussions of the missing mass, motions of galaxies within clusters and the unusual spins of spiral disks do suggest there is additional stuff filling the universe—dark and hidden material ten times more plentiful than luminous matter. But, if Guth is right, another tenfold increase is required to flatten the universe out entirely. This is why the search for exotic particle candidates is so attractive to cosmologists.

By the start of the new year—1980—Guth experienced another stroke of luck. While having lunch at the accelerator laboratory's cafeteria, he heard a colleague, physicist Marvin Weinstein, mention another longstanding cosmological mystery: astronomy's difficulty in explaining the incredible uniformity of the universe. Guth, the cosmology novice, had been unaware of this problem, but after listening to the lunchtime discourse he soon realized that inflation could easily provide an explanation.

It may seem odd to describe the universe as uniform. The presence of stars and galaxies makes it appear quite the opposite. From our earthly perspective, the cosmos looks rather lumpy. We're not only circling the outer perimeter of a clump known as the Milky Way, we're also perched at the edge of that disklike assemblage of galaxies called the Local Supercluster. Even farther out, galaxies appear to surround empty space, to be arranged on the surfaces of immense bubbles. But if you could look down at the universe over scales of billions of light-years, you'd see that the galaxies are fairly smoothly distributed, not unlike the way the foamy texture of a sponge would look increasingly uniform as you stepped away from it.

Under the standard model of the Big Bang, there is no easy way to explain this smoothness. There is not enough time during the early stages of the explosion to get all bits of matter "well blended" before they shoot off. Yet, somehow, the galaxies in the northern part of the sky went on to form and develop in exactly the same way that they evolved in the south. And that background hum of microwaves in our universe, the echo of the Big Bang, varies from place to place in the sky by only one part in ten thousand. However, all regions of our ever-expanding cosmos could not possibly have communicated with one another (in a sense, gotten their story straight) at the time that radiation was emitted. Then, why is each corner of the universe putting out precisely the same hum?

Inflation could be the reason. Recall that right before inflation took off, the region that was going to evolve into our observable universe was a trillion times smaller than a proton. Thus, it was quite easy for all corners of this infinitesimal speck to readily mix, attaining the same temperature and density. Inflation then stepped in to spread and maintain this uniform mixture throughout a growing bubble of space-time.

Like any new scientific venture, though, Guth's original idea was not without its shortcomings. For one, he could not provide a graceful exit from the inflationary spurt. At first, he thought the hyperacceleration might end suddenly, like a runaway car slamming into a brick wall. But that, in theory, left him with a chaotic mess of tiny "bubble" universes, none of which could grow and evolve into the universe we see around us. The crystallization, in a sense, was patchy. Guth joined forces with Columbia University's Erick Weinberg to see if they could come up with a way to get those bubbles of "normal space" to somehow coalesce and form one big universe, but to no avail.

The death knell was about to sound for Guth's inflationary scheme when several other physicists, Linde of the Lebedev Physical Institute in Moscow and, independently, Andreas Albrecht and Paul Steinhardt of the University of Pennsylvania, came to the rescue. In a move that would make Madison Avenue proud, their revised model came to be called the "New Inflationary Universe." By tweaking some of the parameters in the equations, Linde, Albrecht, and Steinhardt were able to show how inflation could proceed in a subtly different manner, lasting long enough for any one of Guth's many bubbles to balloon into a suitable cosmos. In fact, the new inflationary model predicts that the visible universe we observe out to the farthest quasar is just a tiny fraction of the space-time domain that burst forth to give birth to us. The equations can't tell just how much larger, but a good guess is that the entire region is billions and billions of times bigger than the observable universe. If true, astronomers will no longer be able to glibly talk about seeing out to the edge of the cosmos. The best telescopes today can observe quasars, the luminous cores of young, violent galaxies, out to a distance of 12 billion light-years. But the true boundaries of the space-time domain that inflated may now lie more than 10^{30} light-years away—100 billion billion times more distant.

Did it happen just once? Maybe not. The new inflationary scheme also enables physicists to imagine a host of universes frothing out of a primordial sea of symmetry. Some theorists envision these cosmic cousins nesting together like a string of suds blown out of a bubble blower; in other versions, they remain forever isolated from one another. Either vision nicely parallels the debate astronomers had almost a century ago on whether our Milky Way was just one of many "island universes" floating in outer space. From a philosophical standpoint, it almost appears to be the next logical extension: We are but one of many planets circling one of many stars that resides in one of many galaxies . . . in one of many universes. As Peebles would declare, "The inflationary universe allows the imagination to roam free. Perhaps you start matter roaring around, collapsing and expanding. Somewhere you enter an inflationary phase, and it all happens. From this chaos can pop up this and other universes."

Currently, theorists are attempting to wring one more useful function out of the inflationary universe theory: explaining the origin of galaxies. Inflation is quite successful in explaining the overall uniformity of the heavens, but astronomers can't easily forget that as they examine tinier and tinier slices, clumps like galaxies and clusters start emerging. How did a universe so uniform on the largest scale, they must ask themselves, get so mottled on the small scale? It has long been assumed that the Big Bang sent a series of waves coursing through the newly born sea of particles and that these density perturbations pushed and squeezed the gas into galaxies and clusters.

How did these waves originate? No one knows for sure. But inflation provides a good guess. It suggests that the ripples may have been born when quantum fluctuations in that initial kernel about to inflate—submicroscopic disturbances in its sea of symmetry—were blown up to an astronomical scale as the universe ballooned outward. It would be these perturbations that eventually corralled the primordial gas into clumps. Unfortunately, the strength of this effect is very sensitive to the particle physics theory being plugged into the equations. When theorists used the simplest grand unified theory, galaxies tended to collapse into black holes at a rather early point in the universe's history, a result that soon sent everyone scurrying back to the blackboard. But the perfect grand unified theory has yet to be devised. Inserting the proper one into the inflationary scheme, it is hoped, will show how one can form galaxies just like our own. Some believe, in fact, that it will be the key to developing the correct theory.

Predicting how our universe behaved at these incomprehensible times, the first trillionth of a trillionth of a trillionth of a second, seems an audacious endeavor. Guth himself thought so just a few years ago. But inflation's successful track record in solving those cosmic dilemmas —flatness, uniformity, and the dearth of monopoles—can easily change one's mind, and one's status. Guth was finally able to put his gypsy days behind him; he's now a tenured professor in the theoretical physics department at MIT, his alma mater. "Actually, predicting the early universe's behavior is a lot easier than you might expect," says Guth. "The more you heat a system and the hotter it gets, the simpler the interactions. And when you measure the temperature of the microwave background radiation, you find it's uniform to an extraordinary degree —one part in ten thousand. Meteorologists could make terrific predictions if they could deal with climates that uniform. So, in a way, predicting the state of the early universe is a lot easier than predicting the weather!"

What Guth has not told us is this: Where did his initial kernel of mass-energy come from? The very instant of creation remains unexplained. He merely assumes, as all Big Bang theories have up to this point, that our cosmos emerged from a sort of cosmic egg that was exceptionally hot and expanding. We may never have conclusive proof as to the exact nature of that primordial seed, for as Guth puts it, "Inflation simply wipes out all the evidence." But physicists are inveterate dreamers whose hypotheses, speculations, and even hunches have a way of breaking down what seem like insurmountable barriers. Even now, as we shall see in the next chapter, serious physicists are beginning to ponder what was once deemed a ludicrous question: "What was happening," they are asking, "*before* the Big Bang?"

An astronomical rendering of eternity. A long-term photographic exposure, taken outside the dome of the Anglo-Australian Observatory in New South Wales, depicts star trails around the south celestial pole. *Anglo Australian Telescope Board, 1977. Photograph by David Malin.*

11

BEFORE THE BEGINNING?

In answer to the question of why it happened, I offer the modest proposal that our Universe is simply one of those things which happen from time to time.

—Edward P. Tryon

Versions of the tale can be found in many religions and philosophies, but it is sometimes associated with a seventeenth-century Irish theologian named James Ussher. Bishop Ussher was an influential authority on biblical chronology, who spent many years poring over the Bible to pinpoint the exact moment of creation. Even many scientists of the time, constrained by orthodox Christian concepts, viewed this exercise as a valid means of determining the age of the cosmos. Sir Isaac Newton himself calculated a date for genesis in his *Chronologies of the Ancient Kingdoms.*

But it was Ussher, as the legend goes, who was faced with cosmology's greatest enigma. When the bishop solemnly announced to his fellow churchmen that God began to forge the heavens and the Earth at 2:30 in the afternoon on Sunday, October 23, in the year 4004 B.C., one brave soul is alleged to have asked, "And pray, Holy Father, what was God doing *before* he created the universe?" To which Ussher thunderously replied, "Creating Hell for those who ask questions such as that!"

Today, this dilemma remains just as hellish. Each advance that we have witnessed in cosmology these last few decades has increased our understanding of how the cosmos acted at earlier and earlier epochs. As noted before, cosmologists are like archeologists, who dig ever deeper into the strata to uncover the fragmentary shards of a civilization down to the very moment of its origin.

Continuing with this analogy, we might think of Alan Guth as having taken us down to a level marked "10^{-35} second." But what was the condition of the universe before that time? That's a very difficult question to answer. When theorists run the cataclysmic eruption further and further back—like some kind of motion picture film in reverse—every bit of mass-energy ends up packed into an infinitesimally small speck.

This outcome leads to questions as bothersome as the one posed by Ussher's follower: How did that primordial egg of seemingly infinite density and infinite temperature get there in the first place, and why did it explode? In other words, how do you get *something* out of *nothing* without invoking divine intervention?

For many years, this problem was handily sidestepped by the suggestion that our universe oscillates. In this model, the expansion that our cosmos is now experiencing is merely one in an endless sequence of expansions and contractions. This puts the question of our origins in the infinite past, where science doesn't have to worry much about it. What preceded the Big Bang was a never-ending cycle of birth, death, and rebirth. The relentless pull of gravity gradually halts each expansion and draws the galaxies inward until they reform that primeval fireball from which they emerged. And like some bouncing ball, the universe rebounds into another burst of expansion, forming a new generation of galaxies, stars, and planets. Yet many theorists are convinced that a cyclic universe, if it exists at all, will always run down, much the way a bouncing ball eventually peters out. Except that in this case, each bounce, strangely enough, takes longer than the last one, which means only a limited number of cycles could have preceded our present bounce. No longer would the universe have an infinite past, and this conclusion thrusts the problem of our beginnings back into the spotlight.

Some Big Bang cosmologists simply ignore the perplexity by declaring that any comments on the conditions that might have existed prior to the explosion will always remain outside the domain of physical science. Stephen Hawking summed up this viewpoint by noting that "time ceases to be well defined in the very early universe just as the direction 'north' ceases to be well defined at the North Pole of the Earth. Asking what happens before the Big Bang is like asking for a point one kilometer north of the North Pole." But Hawking himself, a physicist who sits in the Lucasian Chair of Mathematics at Cambridge, as Newton did, is far from discouraged by the conundrum. He has also stated that "the initial conditions of the universe are as suitable a subject for scientific study and theory as are the local physical laws. We shall not have a complete theory until we can do more than merely say that 'things are as they are because they were as they were.' "

Physicist Edward Tryon of New York City's Hunter College believes part of the problem has been psychological. "The universe overwhelms one's psyche with its size, and that poses a psychological barrier to people seriously trying to ask, 'How could the universe have emerged following known physical laws?' "

But in recent years these formidable psychological walls have begun to crumble. A few theorists are daring to wrestle with that once unthinkable question: pondering what was happening at "time zero." Their musings are not vague speculations hashed out over coffee in a physics department lounge, but rigorous mathematical treatises which adhere to the laws of particle physics, quantum mechanics, and Einstein's theories of relativity. The titles of the scientific papers themselves display a certain boldness. One enigmatically suggests the "Creation of Open Universes from de Sitter Space"; another more brazenly hawks the "Creation of Universes from Nothing."

Many of these venturesome physicists admit that the details of their models may appear charmingly quaint in coming years as the laws governing the actions of highly energetic subatomic particles are better understood. But their intriguing (some may say disturbing) conclusions may not change: Building on the picture painted by Guth, their equations are telling us that our cosmos may be only one of many universes bubbling out of a yet-to-be defined dimension. Each appears as a sudden *pooff* out of the void. It is a vision of genesis that is sure to knock some philosophical underpinnings out from under us, wrenching askew our place in the cosmos. As Tryon so wryly stated in a paper in the British journal *Nature,* "Our Universe is simply one of those things which happen from time to time."

Tryon himself planted the seed for these contemplations in the late 1960s when, as an eager young assistant professor at Columbia University, he attended a talk being given by a prominent British cosmologist: "I was sitting off to the right side of the room hazily listening when, during a pause in the lecture, I suddenly blurted out, 'Maybe the universe is a vacuum fluctuation.' " By vacuum fluctuation, he meant an instantaneous blip out of nothingness. "But I was crestfallen," he recalls, "when everyone in the room burst into laughter. With three Nobel laureates in the audience, I wasn't about to tell them I wasn't joking. I mean, who was I to tell them where the universe came from?"

But the idea came back to haunt him a few years later as he was preparing a popular article on cosmology. He was captivated by the fact that astronomy cannot yet determine with absolute certainty whether our universe is open or closed. As we saw in previous chapters, this depends on the density of matter in the cosmos, and the amount of gravitating mass measured or conjectured so far lies relatively close to the turning point. In cosmology, being within 10 percent, at the very least, is close.

Tryon speculated on what might be maintaining this precarious near-equilibrium. "We learned from Einstein in his famous equation $E = mc^2$ that matter is a form of energy, and our universe contains an enormous amount of matter," he points out. "But there is also another form of energy important to cosmology that acts, in some sense, in opposition to this mass-energy. Namely, gravitational potential energy." This is the gravitational energy that each star, planet, and wisp of gas possesses by virtue of interacting with every other bit of mass in the heavens. One could think of this as the supply of energy needed to push the galaxies infinitely far apart; hence it is traditionally regarded as a negative energy on the ledger books of the universe.

Maybe the enormous amount of *positive* mass-energy in the universe, Tryon surmised, is perfectly balanced with the *negative* gravitational potential energy present in the cosmos. In a universe inflated to the brink of closure, as ours might be, these energies would cancel each other out in a sort of stalemated tug-of-war. This suggests that the total energy of our universe is actually zero! "If this be the case," Tryon wrote in *Nature* in 1973, "then our Universe could have appeared from nowhere without violating any conservation laws." The universe could indeed appear out of nowhere if it is nothing after all.

Tryon says he was immediately intrigued by this concept because "it had a cold, precise, impersonal beauty about it." The idea of a zero-energy universe does mesh nicely with other conservation laws in our cosmos. For every positively charged particle floating in space, for example, there seems to be a negatively charged particle around to neutralize it, resulting in a net electric charge of zero. And for every galaxy receding from us in the northern hemisphere, there's a corresponding galaxy racing away in the southern sector, adding up to a net momentum of zero. "If you're considering energy, momentum, or electric charge, the universe indeed adds up to zero. But if you're interested in theater, art, or fishing, then you might conclude, 'How interesting nothing can be,' " muses Tryon. "Human existence seems to meander in and around these cold, valueless quantities."

What we have not yet considered is how this cosmic teeter-totter originated. Cars and houses never pop out of thin air. Why should a dense kernel of matter and radiant energy that gives birth to an entire universe suddenly materialize, as Tryon suggested? To answer that question we must turn to quantum field theory, the seemingly bizarre set of rules that govern the submicroscopic world of elementary particles and whose origins date back to the first decades of this century.

As physicists ushered in the twentieth century, they were supremely confident that all processes in the universe were precisely calculable as long as one was given enough data. One prospective physics student entering Harvard University around 1900 was even advised to choose another career because in physics, he was told, there was nothing of importance left to be done. This deterministic philosophy had been established two centuries earlier when Newton successfully plotted the orbits of the far-flung planets with his law of gravitation. In one stroke, he proved that an apple falling from a tree and a celestial body moving through the heavens were governed by the same laws of mechanics: The universe ticked on like a mammoth timepiece.

The nineteenth-century French mathematician Laplace was one of determinism's most enthusiastic proponents. He ventured that if some all-knowing intelligence could observe all the forces acting in nature and record the momentary positions of every bit of matter at any given instant of time, then it "would be able to comprehend the motions of the largest bodies of the world and those of the smallest atoms in one single formula . . . nothing would be uncertain, both future and past would be present before its eyes."

But by the time the Victorian era came to a close, this smug assurance vanished as physicists tried to apply these mechanistic laws to the workings of an atom, an entity a billion times smaller than a golf ball. Events in this minuscule realm do not flow smoothly and gradually with time; rather, they change abruptly and discontinuously. Atoms, for example, can absorb or emit energy only in discrete packets called quanta (hence the term *quantum* mechanics). Nature at this level operates, not as a machine, but like a game of probability. Throughout the first few decades of this century, scientists came to realize that atomic particles behaved with less predictability than such ordinary objects as cars or pencils. The words "always" and "never," used so freely in describing processes in the macroscopic world, were replaced with the terms "usually" and "seldom." Nothing could be counted on, nothing ruled out.

Quantum mechanics' equations of probability even predict that atomic particles can turn up in places where classical laws declare they cannot be. Take radioactivity. By all the laws of nineteenth-century physics, there just isn't enough energy around for a proton or neutron to break free from a nucleus's grip. Yet, the *click, click, click* of a Geiger counter stationed near a piece of uranium instructs us that some particles obviously escape from time to time. Even in the strange world of modern physics, this is quite an amazing feat—as if you had a car locked in a garage and suddenly found that it could tunnel through the garage door to the driveway.

By 1927, German physicist Werner Heisenberg keenly grasped that the probabilistic nature of the laws of quantum mechanics places subtle limitations on how much we can ever know about an atomic system.

One might expect the state of an atom to be characterized completely by the position and velocity of its constituent particles, as well as by its energy. But Heisenberg declared that these quantities would always remain uncertain to some degree. Appropriately enough, this limitation came to be known as Heisenberg's uncertainty principle. It stated that the more accurately you pinpoint the position of a particle, the less accurately you will be able to predict its velocity, and vice versa. This quandary is somewhat intuitive when you think about the equipment needed to make the measurement. Measuring devices are so large and the particles so small that the very act of measuring one parameter is bound to change the other. For scientists used to calculating the motions of projectiles and planets with great exactitude, this was distressing news. Einstein himself railed against this uncertainty and announced that he "shall never believe that God plays dice with the world." It is probably the statement more often quoted than any other from Einstein's long career.

Eventually, though, Heisenberg's uncertainty principle would affect scientists' thoughts about the most sacred law of physics: conservation of matter and energy. In our workaday world, matter and energy can be neither created nor destroyed. Gasoline never materializes in our gas tanks (unfortunately), and when you burn one gallon of the fuel, you end up with exactly one gallon's worth of power and residue. But on the atomic scale, there's a loophole in that law. That tiny degree of uncertainty that Heisenberg told us would always exist at submicroscopic levels allows for minor fluctuations in the energy of a system over very brief moments of time.

For instance, within one billionth of a trillionth of a second, an electron and its antimatter mate, the positron, can emerge out of nothingness without warning, come back together again, and then vanish. This is more than mere speculation; the effects of these spontaneous acts of creation and annihilation have been measured in the laboratory, in precise agreement with theory. Physicists are helpless in explaining exactly how this happens; they know only that it's one possible outcome as nature keeps throwing down its dice. Conservation laws do not forbid it, and quantum-mechanical uncertainties make it inevitable. As Nobel laureate Murray Gell-Mann once remarked (paraphrasing a statement in

T. H. White's *The Once and Future King*), "Everything that is not forbidden is compulsory."

Such unusual goings-on at the subatomic level have given physicists a new perspective in their understanding of empty space. To physicist Heinz Pagels of Rockefeller University, the vacuum comes to resemble the surface of the sea: "Imagine that you're flying over the ocean in a jet plane. From that vantage point, the sea looks perfectly smooth and empty. But you know, when you get down close to it in a small boat, that huge waves are fluctuating all over the place. Well, that's the way it is with empty space. Over large distances—the scales that we experience as human beings—space appears completely empty. But if you were able to probe it very closely, you'd find *all* the quantum particles in existence going in and out of nothingness."

Physicists refer to these ephemeral particles as "vacuum fluctuations." This concept seems to defy common sense, yet is perfectly valid within the framework of quantum mechanics. "No point is more central than this," John A. Wheeler has written, "that empty space is not empty. It is the seat of the most violent physics."

And that was the key to Tryon's 1973 version of the Big Bang. In Heisenberg's uncertainty principle, energy and time are related in an inversely proportional way, like a seesaw. The smaller the amount of energy being asked to spontaneously pop into existence, the longer the period of time it can stick around. Tryon took this to the extreme: "I thought that, if the universe actually has zero energy, then maybe it's a vacuum fluctuation of some larger space in which our universe is embedded, a vacuum fluctuation that has persisted for billions of years. I was saying that there was a minuscule probability for 10^{87} particles and photons to spontaneously arise out of the vacuum in one shot." At least one pundit has remarked that this turns Einstein's famous statement a full 180 degrees around. Now, it is dice that are playing God with the universe!

At the time, this struck many people as preposterous, and Tryon was the first to agree that his conjecture was beseiged by problems. Most importantly, it didn't explain in any natural way why the early universe was so hot or why it ended up so big. "It's more likely that a smaller universe would have been created," says Tryon. "Our universe is not only large, it's larger than it needs to be for us to have evolved and found ourselves observing it."

But, at this point, Tryon was only offering up an idea, not a detailed model of the universe's creation. By 1978 Belgian theorists R. Brout, F. Englert, and E. Gunzig attempted to solve this bewildering puzzle with their own creation scenario. Before the Big Bang, they conjectured, there was only flat, empty space. One could picture this in two dimensions as a rubber sheet extending off to infinity in all directions. But eventually, one of those infamous quantum fluctuations would cause this sheet of space to stretch or warp just a tiny bit, making it look like the indentation in a well-used sofa. This curvature in space-time would trigger the creation of particles, which would create more curvature that creates more particles . . . ad infinitum. Like a runaway stampede, this by-the-bootstraps production of matter and energy would finally emerge as a Big Bang, an expanding universe flooded with hot matter.

Yet many physicists were not convinced that flat, empty space could be so unstable. The Belgians also had to assume that this creationary factory eventually stopped production; but why doesn't this cooperative process just keep going on forever? What was lacking at this stage was a full understanding of the Big Bang itself: knowing how matter, energy, and space-time behaved under the terrifically high temperatures existing in the first 10^{-35} second of our birth. Fortunately, this situation was changing. Even as Brout, Englert, and Gunzig were writing down their equations, particle physicists were starting to appreciate the usefulness of applying their latest theories to the Big Bang in order to unravel its dynamical history. As we saw in the preceding chapter, this effort led to Guth's conjecture that the early universe underwent a short burst of superaccelerated expansion.

Inflation became the answer to a universe creator's prayer. "Alan Guth taught us how to make something big out of something very little," points out Tufts University physicist Alexander Vilenkin. And this lesson enabled new "creationists" to pick up where Tryon left off. Recall that Tryon talked of a vast fountain of particles spontaneously arising out of the vacuum of empty space. But that was never really tenable to the physics community at large, especially after a far more likely scenario was forged. Once the inflationary model took center stage, the quest for our origins was couched in new terms: How to bring into existence that amazing bubble of supercooled symmetry just bursting with energy and ready to inflate. Theorists came up with a number of possible routes.

Pagels, in collaboration with David Atkatz of Bell Laboratories, had an interesting guess. The two physicists ventured that our fetal universe might have resembled a radioactive atom. Perhaps a primeval mini-universe billions of times smaller than a hydrogen atom, seemingly stable, suddenly turned into that inflationary bomb, much the way certain atoms can unexpectedly transmute into different elements by shooting off protons, neutrons, electrons, and gamma rays. "We were struck by the fact," recalls Pagels, "that people once thought that all atomic nuclei were stable. Maybe space is also unstable and has a finite probability of turning into something else," that something else being the Big Bang.

Of course, the question then arises: Where did Pagels and Atkatz's initial pellet of space-time come from? Even Tryon needed his vacuum fluctuation to occur in some empty, primeval space. In a sense, the discovery of our ultimate cosmic roots was merely postponed, not solved. "Space? Time? That's still *something,*" responds Vilenkin in his tiny office on the Tufts campus, blackboards filled with mathematical exotica. The Soviet émigré has more boldly propounded that our universe quantum-mechanically appeared out of, well, nothing.

Denied entrance into Soviet graduate school for political reasons, Vilenkin continued to study and publish on his own in the Soviet Union while working at a series of odd jobs. On one journal paper, his affiliation was even listed as the Kharkov State Zoo in the Ukraine; he was the night watchman at the time. However, Vilenkin's contemplation of "nothing" had to await his emigration to the United States in 1976.

"What do I mean by nothing?" muses Vilenkin. "Nothing is a state without classical space or time, only a space-time foam." The basic notion of a foamy space originated with Wheeler in the 1950s when he attempted to join quantum mechanics with the law that currently best describes gravity for us, Einstein's General Theory of Relativity.

Other forces, such as electromagnetism, always act within the fixed backdrop of space-time. Space and time are mere spectators as the nongravitational forces wield their influence. But gravity is unique, since it is defined as the very distortion of space-time. To have it obey the laws of quantum mechanics requires that space and time be quantized, and thus subject to quantum fluctuations. This means that at some level, as yet undetected, there could be points in space and moments in time that cease to flow smoothly and continuously into the next segment (as if you took one step into your living room and ended up in your bedroom many yards away).

We never experience such discontinuity because it is a process that occurs at levels totally unrelated to everyday physics, even nuclear physics. The smallest unit of length, it is surmised, is around 10^{-33} centimeter. Theoretical physicist Bryce DeWitt of the University of Texas at Austin remarked that this dimension "bears roughly the same relation to nuclear dimensions as the size of a human being bears to that of our galaxy. The unit of time is even more fantastic: 5.36×10^{-44} second. To probe these scales of distance and time experimentally, using instruments built with present technology, one would need a particle accelerator the size of the galaxy!"

Wheeler's guess was that if scientists could explore nature at this unimaginably tiny scale, they'd find the topology of space undergoing incessant and violent fluctuations, similar to the way elementary particles can spontaneously appear and disappear in so-called "empty" space. Only now, it is space-time itself that is popping in and out of existence. Wheeler pictured this occurring at the smallest levels of the space-time continuum that envelops us, but Vilenkin envisioned another possibility: Before the Big Bang, maybe only the foam existed.

According to Vilenkin's vision, there was no classical space, in the traditional sense, to park a star, tree, or subatomic particle. Yet there did exist this ultramicroscopic froth of space-time bubbles. One could imagine this as a cauldron roiling with tiny closed universes (each less than a trillionth of a trillionth of a millimeter across) perpetually growing, collapsing, then disappearing into nothingness. There is no past, no future. Vilenkin's calculations tell us that most of this bubbling out of the void will lead to naught. But there's a small chance that one of the bubbles will suddenly materialize exhibiting all the properties of Guth's inflationary system. Instead of collapsing, this rare power-packed bubble takes off like a rocket, moving explosively outward to release its latent energy. Not only a universe, but time itself is born. This answers that infamous question of why there is something instead of nothing. Nothing is unstable.

It is even possible in this scheme for bubbles to nucleate with differing energies, some leading to universes like ours, some not. For example, one bubble might not have enough power to generate the high temperatures needed to create a bit more matter than antimatter. The equal numbers of particles and antiparticles spawned within this growing bubble might totally annihilate each other, leaving behind a bland sea of photons. That would be an uninteresting cosmos indeed. On the other hand, other escapees from that space-time foam might not settle down to a steady rate of expansion, but instead inflate wildly at hyperaccelerated speeds. "If inflation continues, then there's no chance for even an atom to form, to say nothing of an organism," says Vilenkin. "Only a universe which evolves in the right way allows for the existence of conscious observers."

We were quite lucky. Our universe had the proper conditions so that for every one to ten billion quark/antiquark pairs created and annihilated in the earliest moments of the Big Bang, one extra quark was left over—a bit of the asymmetry that snuck into our universe as the unified force began to differentiate. This tiny surplus was enough to evolve into the stars and galaxies we see today. Like Goldilocks stumbling across the little bear's porridge, our universe seems to have turned out "just right," at least good enough for matter to have clumped into galaxies that produced stars that formed planets that generated biological creatures that could ask such questions of the heavens.

According to a few audacious physicists, our good fortune could be proof that a whole family of sister universes exists. They're disturbed by the fact that our own cosmos would be turned topsy-turvy if certain physical constants were shifted by the tiniest amount. Reduce the strong nuclear force by a few percent, for example, and atoms more complicated than hydrogen would fly apart, making the formation of habitable planets an impossibility. A minute change in the strength of the force of gravity or electromagnetism would alter a star's ability to fuse atoms within the thermonuclear engine at its core, causing the star to either radiate too feebly or burn too quickly for living organisms to evolve nearby. Is it sheer coincidence that nature's basic parameters settled into a range that worked out to our benefit, or is it something more? Some scientists contend this incredible fine-tuning makes more sense if the physical constants were just one assortment of a multitude of possibilities randomly set at the moment of creation. We seemed to have won in this cosmic game of Russian roulette; other universes may not have been so favored, ending up starless and lifeless. According to this "anthropic principle," as it is sometimes labeled, the physical constants must be the way they are, or we wouldn't be around to measure them at all. Our own existence sets constraints on the universe we observe.

The idea that we live in an exceptional universe with specially selected physical constants, it should be stressed, remains sheer speculation; Vilenkin, for one, is very skeptical, since no theory has yet been formulated that can explain how differing physical constants might be produced. He prefers to think that there would be one set of physical laws, one roulette table, upon which any and every universe created plays. In his book *Perfect Symmetry,* Pagels was more adamant: "The influence of the anthropic principle on the development of contemporary cosmological models has been sterile: it has explained nothing; and it has even had a negative influence, as evidenced by the fact that the value of certain constants, such as the ratio of photons to nuclear particles for which anthropic reasoning was once invoked as an explanation, can now be explained by new physical laws. . . . I would opt for rejecting the anthropic principle as needless clutter in the conceptual repertoire of science. . . . The existence of life in the universe is not a selective principle acting upon the laws of nature; rather it is a consequence of them."

At first glance, there seems to be no way to directly prove that we're just a minor bit of flotsam floating within a megauniverse filled with disconnected cosmoses. Each photon, cosmic ray, and galaxy in our universe is forever trapped within the confines of a space-time cocoon, cut off from any opportunity to retrieve a sample of some extra-extraterrestrial material. Yet, an observational test does exist, although it's a real long shot at best. The possibility hinges on a model of creation fashioned by Princeton astrophysicist J. Richard Gott (an ironic surname for an inveterate universe-creator; it's the German word for god).

Sitting amid a jungle of journals and papers in his office overlooking the university's scenic Ivy Lane, Gott compares his work to writing a mystery novel backward. "Particle physicists are trying to write this novel from the first page, hoping it will lead to the proper ending," he says. "But, being an astronomer, I know how it turns out on the last page. So, what I do is ask, 'What kind of creation do we need to produce the universe as it looks today?' "

MIT's Guth taught us that the universe may have undergone a burst of superexplosive expansion that lasted the briefest of moments: 1/100,000,000,000,000,000,000,000,000,000,000 of a second. But Gott takes that idea one step further. He suggests that our normal-looking universe resides in a larger superspace that has been inflating for an indescribably long time and will continue to swell outward for all eternity. In a sense, the inflation that Guth says we experienced during the Big Bang was a momentary peek into this bizarre, ever-expanding netherworld.

Physicists have a name for this peculiar expansionary space. They call it de Sitter space, after the Dutch astronomer Willem de Sitter who first recognized its properties in 1917 while working on a particular solution to Einstein's gravitational field equations. Gott doesn't concern himself with how the de Sitter space got there. "This is because de Sitter space quickly forgets how it got started," he says. "It's very similar to the way a black hole forgets whether it was made out of television sets or old Buicks." Perhaps it is one of Vilenkin's vacuum fluctuations that never stopped inflating.

What's more important, explains Gott, is that de Sitter space is ferociously hot (possibly as high as 10^{31} degrees), exceedingly dense (up to 10^{93} grams of mass-energy in every thimbleful), and markedly unstable. What it wants to do more than anything else is form bubbles of lower density, much the way bubbles of steam will start forming in a pot of boiling water. An observer inside any of these newly forming bubbles would witness first a moment of inflation, then the formation of ordinary matter. Astronomers look back on it all today and call it the Big Bang. Meanwhile, in between the bubbles, there's more and more of this de Sitter space getting bigger and bigger and bigger—roomy enough to contain an infinite number of bubble-universes. As science-fiction writer Arthur C. Clarke once prophetically penned, "Many and strange are the universes that drift like bubbles in the foam upon the River of Time."

This is where the observational test enters: Gott's frothy vision has the potential to turn deadly. The Princeton astrophysicist and his colleague Thomas Statler have determined that there might be discomforting side effects to a megauniverse filled with a bevy of cosmos-sized bubbles. "Since the bubbles are generated randomly in this de Sitter space, some of them will eventually form next to our bubble and hit it," says Gott.

The supracelestial bump could conceivably create an opening between the two universes, starting as a pinpoint but soon growing larger and larger. It might be quite a bright spot in the sky. An opening that began to grow billions of years ago would now stretch half a degree across the sky, about the size of the Moon. Nothing like that has been sighted so far, but this is an event astronomers may not want to detect. The particles that spew into our universe from this interspatial portal could be energized up to a trillion trillion electron volts, as potent as the Big Bang itself! In which case, every solar system, galaxy, and supercluster would soon be vaporized. It would prove that Gott was right, but then no one would be around to congratulate him.

Of course, that is only the most extreme outcome possible; it need not be that dramatic. The momentous meeting might be noticed only as a slight glitch in the microwave background of our universe. This neighborly knock at our space-time door could take place tomorrow, but Gott and Statler's theory currently predicts a very low rate of bubble formation. The collision might not occur for another 10^{500} years (that's a one followed by five hundred zeros, a number almost too enormous to comprehend).

There is no way, in the foreseeable future, to test these varied and intriguing creation scenarios, but even flights of mathematical fancy can play an important role in allowing physicists to peek into the next uncharted territory: a clear understanding of the behavior of the universe when it was less than 10^{-43} second old. This was the momentous period in the universe's history when gravity was allegedly still merged with its sister forces, electromagnetism and the strong and weak nuclear forces, in one *super*grand unified force. From a particle physics perspective, the first 10^{-43} second in the universe's life was the epoch when bosons and fermions, the two classes of particles in the universe, were virtually one and the same, easily flitting from one form to the other.

Many of the exotic candidates for the universe's missing mass spring from this assumption. If a boson in the first instant of creation could change into a fermion (and vice versa), then each boson in existence today should have a unique fermionic superpartner, and likewise each fermion a boson. This is why physicists are seriously considering the possibility that a photon (a boson) is uniquely related to a particle, yet to be detected, called the photino (a fermion), a quark to a squark, a graviton to a gravitino, and so on down the line of elementary particles.

Although they recognize the importance of this quest, young Turks in this business are having a few chuckles over the language this new endeavor has created. An article has been circulating throughout the United States' astrophysical community, for instance, entitled "Superduperstuff in the Universe" by "G. Wow-mann." "Until recently ordinary stuff was widely regarded as sufficient, since any phenomenon could be explained in terms of it by a sufficiently skilled theorist," the bogus tome iterates. "This is not good enough, however. Here we introduce the concept of superconducting, supercolliding, supersymmetric, superstringy superstuff *('Superduperstuff')* which has the attribute that *it possesses all properties whatsoever."*

This is perhaps a not-so-subtle reminder that theories should never be considered sacred and are always vulnerable to being overturned by the next theoretical revolution. "Conceivably," Pagels has cautioned us, "as physicists discover new laws that logically subsume the previous laws, they may find that the process never terminates. Instead of finding an absolute universal law at the bottom of existence, they may find an endless regress of laws, or even worse, total confusion and lawlessness —an outlaw universe. . . . Yet in spite of this possibility, the notion of a simple law describing all existence beckons like the Holy Grail."

Gravity, for one, must assuredly have obeyed the rules of quantum mechanics during that primal moment when all the mass of the observable universe was squeezed into a knot less than 10^{-33} centimeter across. In the past, all the laws of physics have fallen apart as theorists, armed with only paper and pencil, cautiously proceeded into this last frontier of cosmology. Physics will be gnawingly incomplete until a theory of "quantum gravity" is firmly established. Alone, Einstein's theory of gravity "blows up" when it attempts to describe the world at the subatomic level. The only thing that physicists get for their troubles is a basketful of infinities. "Answers that are as absurd," Caltech theoretical physicist John Schwarz has written, "as the results you get if you try to divide a number by zero."

But, by the 1980s, the hope that gravity—a force whose tendrils span the universe—and quantum theory could be fully integrated started skyrocketing. The solution lies in the possibility that the Big Bang was a kind of multidimensional blowout: Perhaps the four dimensions of space and time that surround us today rapidly swelled out of a primeval universe comprised of many dimensions. "This may have a science-fiction ring to it," said Schwarz, "but it is nonetheless a perfectly sensible possibility."

The roots of this idea go back to the time when Einstein had just instructed the scientific community to view gravity, not as a mysterious force of attraction, but rather as a curvature in four-dimensional space-time. Influenced by this geometric description of physical law, a little-known German mathematician named Theodor Kaluza discovered in 1921 that electromagnetism could be described as a warp in yet another spatial dimension—a *fifth* dimension in space-time. Every electromagnetic wave, from radio waves to X rays, could be thought of as a ripple in this extra dimension of space.

By 1926, Swedish physicist Oskar Klein provided a provocative explanation for why we are not aware of this added set of coordinates. He theorized that, at each and every point in space, the fifth dimension is rolled up into a compact cylinder only 10^{-32} centimeter across (a width more than a billion billion billion times smaller than a proton). Therefore, no matter how hard we try, it remains unobservable; the tiny cylinder still looks like a point to our instruments. We have no microscope or particle accelerator big enough to magnify this hidden structure, just as our naked eyes are totally unable to resolve pointlike stars into stellar disks. Our inability to discern the disks, however, makes them no less real.

For decades, the Kaluza-Klein theory was considered little more than a mathematical curiosity, until modern-day physicists realized that its outlandish notion of submicroscopic dimensions could actually help them relate the varied forces of nature under one unified banner. In order to do this, Kaluza-Klein's original scheme was extended to accommodate the two forces discovered since the 1920s, the strong and weak nuclear forces. This demanded that more and more dimensions be added. Currently, specialists in this particular branch of physics imagine a universe woven of ten space-time dimensions, all but four of them ensconced in a realm we cannot perceive directly. Unfortunately, as yet, there is no natural way for physicists to explain why the higher dimensions remained curled up like little balls, while the four we now know as length, width, height, and time jumped out as the Big Bang. Why do only four blow out? Why not six, or eight, or nine? That remains a tantalizing mystery.

Various particle-physics theories involving multidimensional space have been worked on for many years, but a model known as the "superstring theory" has superseded them all. It was the first quantum theory of gravity to get rid of annoying infinities and pesky violations of sacrosanct physical laws without just sweeping them under the rug. At the heart of superstring theory is its redefinition of the fundamental building blocks of nature. At its most basic level, the theory simply states that such entities as quarks, electrons, and neutrinos are not really pointlike particles but rather one-dimensional strings around 10^{-33} centimeter long (not to be confused with cosmic strings). These strings are indeed very tiny: If an atom were blown up to the size of the Milky Way, this string would be no bigger than a human cell.

The most promising version of this theory suggests that the strings come in the form of closed loops. The forces we experience and the particles we detect depend on the manner in which these infinitesimally tiny loops of energy connect, split up, wiggle around, and rotate within a space-time composed of ten dimensions. With this new picture, all the elementary particles we count, collect, and measure within our expansive four-dimensional space-time continuum are merely oscillations in those still-hidden higher dimensions. "Which of the various elementary particles a string represents depends on the manner in which it vibrates," explains Schwarz. "By oscillating in a particular way, for instance, a string would be an electron. Also, strings can join or divide—two coming together to form one or one dividing into two. This interaction is the origin of the fundamental force from which gravity, electromagnetism, and the various nuclear forces are derived." A master of disguise, a single string has the potential to turn into an infinite variety of particles, just as there are innumerable frequencies at which a guitar string can vibrate. We don't see them all because, here in our relatively low-temperature environment, we only get to notice the low-energy states.

The idea that elementary particles could be described as wee bits of strings first occurred in the late 1960s, but it was largely abandoned in favor of other more fashionable mathematical approaches. Schwarz, though, along with several associates, clung to the model with tenacity, especially when they realized that strings showed great promise in explaining a graviton, the hypothetical quantum particle that transmits the force of gravity (just as photons of light transmit the electromagnetic force and gluons carry the strong force).

Such unwavering determination had its rewards. In the summer of 1984, Schwarz and Michael Green of Queen Mary College of the University of London brought strings dramatically back into the spotlight. After a years-long collaboration, the two were able to show that strings could be the long-sought key to the "Theory of Everything"—a way to unite gravity with quantum field theory and consequently the three other forces as well. The dream that had eluded Einstein all his life was finally being fulfilled.

Shadow matter—surely the most exotic solution to the missing-mass problem—is one possible consequence of the superstring theory. The ghostly material shows up when the supergrand unified force undergoes its first breakup. The superstring equations hint that the momentous separation could allow for two completely different sets of particles and nongravitational forces to exist in the same universe. Our strong, weak, and electromagnetic forces go one way; the shadow world's go another. The only force that they share is gravity. There are cosmological reasons, however, to suspect that the shadow world is not exactly parallel to our own. If it were, we would be noticing its effects both on the expansion of the universe and in the way that hydrogen was forged in the Big Bang. But we don't see this. One guess is that shadow matter, if it even exists, is very diffuse. Superstring theory is not remiss, however, in providing additional candidates for the infamous missing mass. As with other theories that have attempted to unify the four forces of nature, superstring theory incorporates the idea that the basic particles now found in the universe do have partners: squarks, photinos, and selectrons may yet be with us.

Some compare the excitement generated by superstring theory to the heady days of the 1920s when quantum theory was first being developed; others, such as Sheldon Glashow, are more cautious, since superstring theory has not yet generated any concrete numbers, such as the mass of a quark or neutrino. In some sense, Schwarz, Green, and the many proponents jumping on the superstring bandwagon with great zeal have only established a blueprint so far, a mathematical means to an end; the entire house is far from being constructed, and tempests could suddenly arise to destroy the foundations.

Construction of mammoth new particle accelerators such as the Superconducting Super Collider (SSC), a proposed United States atom smasher with a 60-mile-long circular tunnel, would assuredly take physicists a few steps closer to the heart of various elementary particles, but not at all near the energies at which superstrings could be seen. For that domain, experimental proofs will have to be indirect. "The existence of the six extra dimensions could be demonstrated by experiments in which particle collisions at high energies produce new kinds of particles with very specific properties," points out Schwarz. "Unfortunately, small dimensions, like small particles, must be investigated in machines that employ enormous amounts of energy. The extra dimensions are probably so small that the energy required to explore them [directly] is far beyond what will be available to use in the foreseeable future." Physicists may have to rely on the simplicity and elegance of the theory to prove its worth.

Some optimists venture to guess that a successful merger of gravity with quantum mechanics will make the origin of our universe self-evident. The indecipherable singularity at the heart of the Big Bang will disappear, they say, perhaps to reveal a state of nature not even imagined as yet by the most far-sighted science-fiction writer. At that time, physicists might be tempted to assume the smug airs of their Victorian predecessors, who mistakenly believed that they had finally unmasked all of nature's secrets. But Hawking, for one, is more realistic. "We are making progress, and there is a reasonable chance that we will discover [a unified theory] by the end of the century," he has said. "At first sight it might appear that this would enable us to predict everything in the universe. However, our powers of prediction would be severely limited, first by the uncertainty principle, which states that certain quantities cannot be exactly predicted but only their probability distribution, and secondly, and even more importantly, by the complexity of the equations, which makes them impossible to solve in any but very simple situations. Thus we would still be a long way from omniscience."

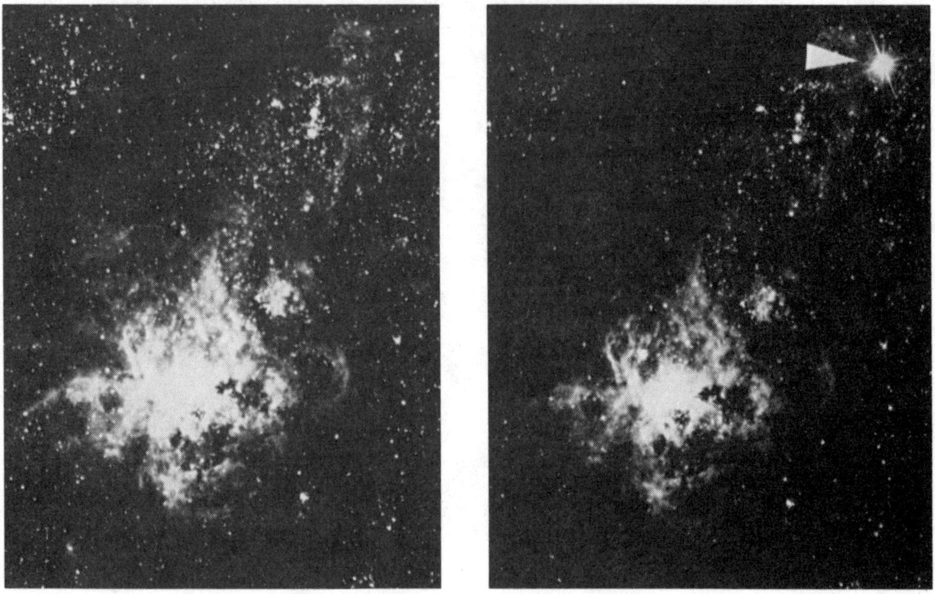

Before and after. The Tarantula nebula in the southern sky (left) was temporarily overshadowed by a visitor (right) in 1987. This guest star was the first nearby supernova visible from Earth in nearly 400 years. By detecting a burst of neutrinos from the blast, astronomers may have witnessed the birth of a neutron star—an historic first. *Courtesy of the National Optical Astronomy Observatories.*

EPILOGUE

> My next reaction was, "That's one hell of a flaw."
> I knew there couldn't have been a flaw that big.
>
> —Ian Shelton, describing his first glimpse of Supernova 1987A on a photographic plate.

Writing about astronomy's frontier is risky because its borders can never be clearly delineated. The boundaries expand yearly—sometimes subtly, sometimes abruptly. New stars and planetary systems are discovered; new theories are expounded; and new glitches appear in telescopic detectors, awaiting an explanation. By the twenty-first century, a gravity-wave telescope, watching for undulations in the fabric of space-time, could well register some blips as momentous as the ones sent out by Jocelyn Bell Burnell's "Little Green Men," altering our cosmic perspective once again. Indeed, we are now entering an era in astronomy that steps beyond the electromagnetic spectrum as a source of information on the universe-at-large. Decades from now, science historians may decide that this new epoch was inaugurated in the winter of 1987 when underground detectors, originally built to observe protons decaying, recorded an onslaught of ghostly neutrinos spewed from the heart of a giant stellar explosion. With this historic particle burst, researchers may have witnessed the actual formation of a neutron star—and by a means far removed from the lenses, mirrors, and antennae of conventional astronomy. The birth of a supernova was once a theoretical game, played out on computers. Now, it has been transformed into a science based on real observations. Anticipated for nearly four hundred years, the supernova that visibly blossomed in the southern sky became the astronomical event of the 1980s. A dozen generations of astronomers have lived and died without ever having had the chance to watch a nearby star explode. This generation is more fortunate, and it is making the most of the opportunity.

The momentous saga began in the early morning hours of February 24 when Ian Shelton, graduate-school dropout and resident technician for a University of Toronto telescope at the Las Campanas Observatory in the Andes mountains of northern Chile, stumbled upon what ancient Chinese astronomers would have called a "guest star." Examining a routine photograph he had just made of the Large Magellanic Cloud, a small companion galaxy of the Milky Way, Shelton noticed a curious blotch. Stepping outside to check the dark Chilean sky for himself, he saw that the pinpoint of light was less bright than the stars in the Big Dipper but brighter by far than anything else in the Magellanic Cloud, which is normally viewed from the southern hemisphere as a mistlike haze. More important, the mysterious object hadn't been there the night before. Oscar Duhalde, night assistant on another telescope at Las Campanas, had also noticed the new star while on his way to a midnight coffee break. Sparked by Shelton's announcement, Duhalde and a colleague drove to the nearby town of La Serena and quickly dispatched a telegram to the Central Bureau for Astronomical Telegrams in Massachusetts, the world's arbiter for astronomical discoveries. Confirmed by additional sightings from New Zealand and Australia, the guest star was officially designated Supernova 1987A, for it was the first stellar explosion that was sighted that year.

Word spread fast, and soon an arsenal of space- and ground-based equipment was aimed at the brightening spot: radio dishes in South Africa and Australia; x-ray and gamma-ray detectors aboard satellites; airborne infrared instruments; and optical telescopes throughout the southern hemisphere. Excitement over the explosion was understandable: Modern-day astronomers have observed hundreds of exploding stars in other galaxies situated millions of light-years away, but the great distances involved prohibited any thorough examinations. Johannes Kepler spotted the last visible supernova in our galactic neighborhood in 1604, before the invention of the telescope. But Supernova 1987A, located only 160,000 light-years from Earth, changed all that. Such a relatively close explosion serves as a golden opportunity, centuries overdue, to confirm or challenge current theories on a massive star's demise. Certain long-standing notions concerning stellar explosions have already been modified because of this supernova's quirky conduct.

During the first few days of the explosion, the supernova's brightness increased a thousandfold, much swifter than was expected, only to plateau at a level far dimmer than predicted. And although very hot and blue initially, the supernova turned red astonishingly quickly, a change that occurred as the shell of debris raced outward at 50,000,000 miles per hour and cooled. In March, the supernova brightened once again, this time powered by the decay of radioactive elements in the stellar remains. A vast quantity of nickel and cobalt, forged in the detonation, was decaying into iron—enough iron to construct 20,000 Earths. At its peak, toward the end of May, the Magellanic blast was as luminous as billions of Suns, but curiously less brilliant than stellar explosions observed in other galaxies. As the summer of 1987 commenced, the supernova began a slow decline into obscurity. A year after Shelton's initial sighting, the visible fireworks were completely over; to the unaided eye, Supernova 1987A had faded into oblivion in the southern sky.

Understanding the supernova's peculiar ups and downs was difficult at first because astronomers were led astray in identifying the progenitor star. For a star to blow up instead of quietly flaming out, its mass must be at least eight times greater than our Sun's. After carefully examining preblast photographs of the Large Magellanic Cloud, observers chose a blue supergiant star, known as Sanduleak −69° 202 in a star catalog prepared by astronomer Nicholas Sanduleak, as the most likely victim. But Harvard astronomer Robert Kirshner, utilizing the International Ultraviolet Explorer space telescope, assured everyone that Sanduleak-69 was still shining serenely.

The mix-up was temporary. Within two months of the supernova's appearance, Kirshner announced that he and his colleagues had been misled by two other blue stars in the immediate area. "Let's not pussyfoot around," Kirshner said at the time. "Sanduleak-69 was it." Upon hearing this reversal, someone sent Kirshner a can of red herring, and theorists issued a collective sigh of relief.

With the host star pinned down, many pieces of the celestial puzzle began to fit. As pointed out in an earlier chapter, astronomers have long surmised that the most likely type of star to go supernova is a red supergiant, a blue star that turned ruddy as it swelled to gigantic proportions in its old age. But the explosion of a smaller, though still massive, blue giant star explains much of the Magellanic supernova's unusual behavior. Sanduleak-69 was a little less spectacular in its death throes simply because the blue star's more compact supply of material both dampened and hastened the visual display. Astronomers now suspect that a sizable fraction of exploding stars are blue supergiants and that these relatively dim supernovae have been overlooked in the past simply because they are hard to see, especially in far-off galaxies. The Magellanic supernova isn't rewriting the textbooks, but it is amending and lengthening them appreciably. Sanduleak-69 may have exploded before it ever advanced to the red-giant stage, or it may have been a red giant that had somehow returned to a blue and smaller state, perhaps by having some of its outer layers waft away in a steady stellar wind over hundreds of thousands of years.

In any case, computer simulations by stellar-evolution experts, such as Stan Woosley of the University of California at Santa

Cruz, suggest that Sanduleak-69 lived fast and died young. Born just ten million years ago, the blue supergiant was about 20 times as massive as our Sun, 40 times larger, and 100,000 times more luminous. It also aged 1,000 times faster. Over those millions of years, as hominids on Earth were learning to walk upright, the nuclear furnace in Sanduleak-69's center generated energy by fusing matter into ever-heavier elements. First, hydrogen was converted into helium; then helium fused into carbon and oxygen, which served as the raw materials for even heavier elements, such as magnesium, silicon, sulfur, and calcium. Once iron was formed, though, the game was over: The fusion of iron in a stellar core uses more energy than it releases.

Without fusion-generated heat, the Mars-sized iron core could not withstand the force of its own gravitation. In less than a second, it collapsed into a sphere about a dozen miles in diameter, but 1.4 times as massive as the Sun. All the protons and electrons merged to form an ultradense ball of neutrons, instantly creating a firestorm of 10^{58} neutrinos that sped out of the star in all directions at or near the speed of light, a titanic prelude to the visible explosion to come. Ninety-nine percent of the supernova's power—roughly equivalent to converting a tenth of our Sun's mass to pure energy—was carried off by these neutrinos. During that split second, Sanduleak-69 released more energy than the combined optical radiation of all the other stars in the universe. The light given off by the supernova later—bright as it was—was surprisingly incidental.

Like a compressed coil, the newly squeezed core rebounded a bit, generating a powerful shock wave that spent a couple of hours working its way through the rest of the star. Light elements were fused into heavier ones—even elements beyond iron—until the star's outer envelope was completely blown off, emitting the initial burst of light that traveled some 160,000 years before arriving at Earth. The detonation scattered the stuff of future stars and planets—from carbon and oxygen to calcium and uranium—into space. By spreading its ashes in this way, the supernova became a driving force behind the chemical evolution of the universe. Only a neutron star, feverishly twirling on its axis, likely remains behind at the exact site of the explosion. If the neutron star's orientation is favorable, detectors may register a periodic "blip" as the pulsar whirls around, causing beams of radiation, from X rays to radio waves, to sweep regularly across the Earth.

Because of a stunning measurement, an observation unprecedented in the annals of astronomy, astronomers are fairly certain that the neutron star is there: The exact moment of the neutron star's formation was unwittingly recorded nearly a day before the luminous explosion was first glimpsed from the Andes mountains. Upon hearing of the supernova's appearance, a number of theorists around the world quickly began to calculate the number of supernova-generated neutrinos that might have been captured on Earth with special instruments—huge tanks of water set in underground mines far from disruptive cosmic rays. When researchers at various sites sifted through their data banks, two groups did indeed find the predicted burst in their computer records.

An individual neutrino stops for almost nothing; it can pass through trillions of miles of matter and not collide with a single atom. But scientists know that they can snare a few of the elusive particles out of a passing flood. Of the ten thousand trillion trillion neutrinos that Supernova 1987A shot through the Earth, the Kamiokande II detector in Japan caught eleven during a 13-second span; a similar instrument in an Ohio salt mine, the Irvine-Michigan-Brookhaven detector set 2,000 feet beneath the shore of Lake Erie, registered eight neutrinos. Because both detectors recorded the neutrino signal at virtually the same time—about 20 hours before the visible flaring—astronomers are convinced that the signal was real. Decades of theoretical work on neutron-star formation were confirmed in a matter of seconds—and with remarkable accuracy. "It was mind-blowing," exclaimed Stirling Colgate of the Los Alamos National Laboratory, who was one of the first to point out the vital role of neutrinos in supernova explosions. "For years, we had theory upon theory, like some massive house of cards, trying to figure out the biggest explosion the universe has to offer. Supernova 1987A told us that physical laws can be extrapolated, and then confirmed."

The duration of the supernova's neutrino burst even allowed theorists to "weigh" the particles. Recall that scientists have been trying to answer the question of whether neutrinos have mass or not, an issue with cosmic consequences. Neutrinos are so prolific that their combined weight, if they are more than spots of energy, could cause the universe to stop its expansion and collapse in a Big Crunch. Although an exact mass could not be calculated—the explosion itself smeared the signal—an upper limit on the mass of one type of neutrino was obtained: The mass of this neutrino appears to be too small to affect the universe's evolution—"well below the mass needed to close the universe back up," concludes John Bahcall of the Institute for Advanced Study in Princeton, New Jersey.

With the death of a star came the birth of a new science, boosting plans for the construction of even larger neutrino "telescopes." Of all the measurements made of the Magellanic supernova, the neutrino detection—the first of its kind—will surely be the achievement best remembered. Not that the supernova itself is in any danger of being forgotten by scientists. "Although a supernova catches the public's imagination while it is brightest," points out Woosley, "it's actually more interesting as it fades." As the supernova expanded, it became more and more transparent. Consequently, astronomers were able to probe deeper and deeper into its inner layers. By the end of 1987, the remnant had spread over hundreds of billions of miles. By then the stellar debris was thin enough for high-energy radiations to shine through, allowing scientists to gather the first direct evidence that supernovae are indeed fiery cauldrons generating the elements essential to life. By conducting "autopsies" on centuries-old supernova remnants and by carrying out painstaking nuclear measurements in the laboratory, astronomers had deduced that a large fraction of the basic elements in the universe was manufactured in massive stars and launched into space when the stars exploded. Elements heavier than iron, it was presumed, were largely created at the moment of the cataclysm. Evidence for these assumptions, however, was always circumstantial until Supernova 1987A.

A gamma-ray detector aboard the Solar Maximum Mission satellite, in Earth orbit since 1980 to study the Sun, began to see gamma rays emanating from the supernova in August 1987. Scientists at the Naval Research Laboratory in Washington, D.C., determined that the gamma-ray energies were precisely those expected when radioactive cobalt slowly decays into iron. Presumably, huge amounts of nickel forged in the blast had earlier decayed into the cobalt. The Solar Max signal was confirmed later that fall when two research teams mounted gamma-ray detectors aboard balloons and lofted the sensitive instruments from central Australia to an altitude of 120,000 feet. To be successful, the detectors had to rise above much of the atmosphere, which absorbs gamma rays. "These detections were the culmination of 20 to 30 years of effort," says Gerald Share of the Naval Research Laboratory, "proving that supernovae are the breeding ground for elements from iron to uranium and that such cataclysmic stellar deaths planted the seeds for the birth of life on Earth." The iron in our blood, in other words, could consist of atoms that were fused in massive stars that exploded billions of years ago. Supernova 1987A had directly confirmed the long-held theory of explosive nucleosynthesis.

The story of the Magellanic supernova does not end there. Centuries from today, astronomers will still be studying its wispy, incandescent filaments, just as observers now meticulously examine the remains of the "new stars" observed centuries ago by Tycho Brahe and Johannes Kepler. Theories rise and fall with each new measurement of the ever-expanding remnant. "The supernova has become a Rorschach test," says Woosley, "stimulating the imaginations as well as creative minds of all who contemplate the event."

The trials, tribulations, and triumphs that occurred as scientists chronicled the day-by-day activities of a newborn supernova underscore an unwritten law in astronomy: New vistas are almost always revealed when investigators—observers and theorists alike—quickly respond to the unexpected. This maxim applies to other areas of science as well. Physicists are readying more powerful atom smashers whose multitrillion-electron-volt particle collisions could very well prove that the search for symmetry within the forces of nature, a harmony that would have been apparent only during the first nanosecond of creation, is as simplistic an enterprise as the failed attempt by ancient Greek philosophers to construct the universe out of four elemental components: Earth, Air, Fire, and Water. "I have a secret bet with myself," confesses Heinz Pagels, "that when the next generation of high-energy particle accelerators is built, we're going to discover all sorts of new particles that cannot be explained by any of our current theories. People will then look back on our model-building today and say, 'How naive.' There may be lots of stuff out there, just waiting to be discovered, that radically changes our understanding of the universe."

Scientific models, perhaps like rules, are meant to be broken (or at least amended). "We advance," muses Maryland astronomer Leo Blitz, "because we doubt what has come before." The idea of a cosmos endowed with Aristotelian perfection was no longer viable after Galileo inspected the heavens with his telescope. A few centuries later, the mere notion of neutron stars and black holes completely disrupted Arthur Eddington's more modest view of stellar evolution; stars indeed behave in absurd ways. Some of astronomy's current hypotheses are just as vulnerable. It is a Thursday's universe, and we have far to go.

"People often have the general impression that, once a theory is developed, every scientist embraces it like a religion," notes Fermilab's Edward Kolb. "They believe we are not allowed to question the basis of the theory or to think about some other kind of revolutionary model. But that's not true at all. We always desire a devil's advocate to make us think about the consequences of what we are saying. If there were evidence tomorrow that the concept of the Big Bang is wrong, I would be the happiest person around, for it would be a completely new thing on which to work. I would be surprised, but I would not be dismayed."

Discrepancies already loom on the horizon. A gigantic tank of cleaning fluid, 100,000 gallons of perchloroethylene, is set nearly a mile beneath the Black Hills of South Dakota to catch a few solar neutrinos out of the hordes that are continually spewed into the solar system from the Sun's nuclear furnace. Every second, 66 billion solar neutrinos are believed to strike each square centimeter of the Earth's surface. Chlorine atoms in the fluid are able to capture about one of the evasive particles every three days, but this is roughly a third of the number expected to be caught. Many theories have been brought forward to account for the discrepancy, but to some it implies that the rate of nuclear reactions inside the Sun is far less than the rate required by theory to maintain the Sun's present luminosity. "Either astronomers don't know how the Sun shines or physicists don't know how neutrinos propagate," says John Bahcall, who has calculated the Sun's neutrino output. "If astronomers are wrong, it is a very serious matter. Many concepts in astronomy—from the universe's helium content to stellar evolution—will have to be reexamined."

Such unresolved mysteries, however, cannot in any way diminish the many accomplishments that have already been made. A small planet in the outer reaches of a galaxy may seem insignificant when balanced against the immensity of the heavens. But more awe-inspiring is the ability of men and women to investigate mere spots of light, shimmering in the nighttime sky, and from their examinations arrive at such a magnificent description of creation. "In the last analysis," it was noted in 1948 during the opening of the famous Hale telescope, "the mind which encompasses the universe is more marvelous than the universe which encompasses the mind."

March 1988

ACKNOWLEDGMENTS

It was in writing this book that John Donne's memorable phrase "no man is an island" took on true meaning for me. During the research phase of this endeavor, I traveled across the United States several times, speaking to more than one hundred scientists. Space prevents me from thanking each separately, but nearly all who were directly quoted from my interviews very graciously agreed to review the sections that involved their work. They not only pointed out my errors but often made many beneficial suggestions as well. I am grateful for their insights. Any mistakes that remain in the preceding pages are totally of my own doing.

Thanks must also be given to those astronomers, theorists, and engineers who took the time to provide me with invaluable background material regarding their fields of interest. They include David Batuski, Kenneth Brecher, Thomas Dame, Craig Foltz, F. R. Harnden, Fred Harris, Piet Hut, Burton Jones, John Lacy, Frank Low, Ronald Maddalena, William Mathews, Joseph Miller, Yvonne Pendleton, Joel Primack, Mark Reid, Rudy Schild, Matthew Schneps, Irwin Shapiro, John Stocke, Alex Szalay, Peter Wehinger, Rogier Windhorst, Sidney Woolf, and Erick Young. At the same time, I am beholden to the editors of *Discover*, *Mosaic*, *Omni*, *Science Digest*, *Science 86*, and *Science News*; their assignments acquainted me with many of the topics discussed in this book. Some of the material first appeared in these publications.

For a writer on a tight schedule, arranging a series of interviews or a trip to an observatory can become an administrative nightmare. There were many who smoothed the way. For their assistance, appreciation is due Lauray Yule at the University of Arizona's Steward Observatory,

James Cornell at the Harvard-Smithsonian Center for Astrophysics, Dennis Meredith and Larry Blakeé at Caltech, and Carl Posey and Agnes Paulsen at the Kitt Peak National Observatory. In this regard, a special thank-you is extended to John Gustafson at Lick Observatory (and to his wife Sarah); John's letters were a continual source of inspiration. At Palomar mountain, both Bob Thicksten and Alec Boksenberg were engaging tour guides of the observatory's facilities; Sam Palmer aptly filled the same role during my visit to Columbia University's radio telescope in New York City.

Several close friends and relatives let me hang my hat at their abodes during my travels and kept my spirits high during the course of my stay. For this I thank Elizabeth Maggio in Arizona; Dona Cooper, Dave Stewart, and the Gustafsons in California; L. Cole Smith and Rick Marshall in Charlottesville, Virginia; and the Grussendorfs (Victoria, Mark, Christopher, Andrew, and Elizabeth), then residing in Washington, D.C. I shall not forget their many kindnesses.

I am deeply indebted to Bob Weil at Omni Books for his unwavering encouragement in getting this project off the ground and to Judit Bodnar at Times Books for her superb copy-editing. I offer the greatest appreciation to my editor Hugh O'Neill and his assistant Susan Randol for their ever enthusiastic editorial support.

Lastly, a special note of gratitude must be extended to my parents and to Stephen Lowe . . . for always being there.

BIBLIOGRAPHY

○ ○ ○

●

· GENERAL REFERENCES ·

Abell, George. *Realm of the Universe.* Philadelphia: Saunders College Publishing, 1984.

Alter, Dinsmore, Clarence H. Cleminshaw, and John G. Phillips. *Pictorial Astronomy.* New York: Harper & Row, 1983.

Bok, Bart J., and Priscilla F. Bok. *The Milky Way.* Cambridge: Harvard University Press, 1974.

Brandt, John C., and Stephen P. Maran (editors). *The New Astronomy and Space Science Reader.* San Francisco: W. H. Freeman and Company, 1977.

Cornell, James, and Paul Gorenstein (editors). *Astronomy from Space.* Cambridge: Smithsonian Astrophysical Observatory, 1983.

Cosmology + 1 (Readings from Scientific American*).* San Francisco: W. H. Freeman and Company, 1977.

Ferris, Timothy. *Galaxies.* San Francisco: Sierra Club Books, 1980.

Friedman, Herbert. *The Amazing Universe.* Washington, D.C.: National Geographic Society, 1975.

Harwit, Martin. *Astrophysical Concepts.* New York: John Wiley & Sons, 1973.

Hoyle, Fred. *Astronomy.* London: Rathbone Books Limited, 1962.

Jastrow, Robert. *Red Giants and White Dwarfs.* New York: W. W. Norton and Company, 1979.

Jastrow, Robert, and Malcolm H. Thompson. *Astronomy: Fundamentals and Frontiers.* New York: John Wiley & Sons, 1977.

Kellermann, K., and B. Sheets (editors). *Serendipitous Discoveries in Radio Astronomy.* Proceedings of Workshop Number 7 held at the National Radio Astronomy Observatory, Green Bank, West Virginia, May 4–6, 1983.

Kraus, John. *Our Cosmic Universe.* Powell, Ohio: Cygnus-Quasar Books, 1980.

Maffei, Paolo. *Monsters in the Sky.* New York: Avon Books, 1981.

Mitton, Simon (editor). *The Cambridge Encyclopaedia of Astronomy.* New York: Crown Publishers, Inc., 1980.

Pagels, Heinz R. *Perfect Symmetry: The Search for the Beginning of Time.* New York: Simon and Schuster, 1985.

Shu, Frank H. *The Physical Universe: An Introduction to Astronomy.* Mill Valley, California: University Science Books, 1982.

Silk, Joseph. *The Big Bang: The Creation and Evolution of the Universe*. San Francisco: W. H. Freeman and Company, 1980.

Tucker, Wallace H. *The Star Splitters: The High Energy Astronomy Observatories.* Washington, D.C.: National Aeronautics and Space Administration, 1984.

· INTRODUCTION ·

Astronomy and Astrophysics for the 1980s: Report of the Astronomy Survey Committee. Washington, D.C.: National Academy Press, 1982.

Gingerich, Owen. "Unlocking the Chemical Secrets of the Cosmos." *Sky & Telescope* (July 1981) 13–15.

Goldberg, Leo. "Prologue: Astronomy Before the Space Age," *Astronomy from Space.* Cambridge: Smithsonian Astrophysical Observatory, 1983, pp. 1–19.

· CHAPTER ONE ·

Blitz, Leo. "Giant Molecular-Cloud Complexes in the Galaxy." *Scientific American* 246 (April 1982) 84–94.

Cohen, R. S., H. Cong, T. M. Dame, and P. Thaddeus. "Molecular Clouds and Galactic Spiral Structure." *The Astrophysical Journal* 239 (July 15, 1980) L53–L56.

Dame, T. M., and P. Thaddeus. "A Wide Latitude CO Survey of Molecular Clouds in the Northern Milky Way," paper presented at The Local Interstellar Medium, I.A.U. Colloquium Number 81, Madison, Wisconsin, June 4–6, 1984.

Eberhart, J. "'Planet' detected beyond the solar system." *Science News* 126 (December 15, 1984) 373.

——— "Vega & Co.: What's Being Born Out There?" *Science News* 124 (August 20, 1983) 116.

Elmegreen, Bruce G. "Molecular Clouds and Star Formation: An Overview," paper presented at Protostars and Planets meeting, Tucson, Arizona, January 1984.

Elmegreen, Bruce G., and Debra Meloy Elmegreen. "Regular strings of H II regions and superclouds in spiral galaxies: clues to the origin of cloudy structure." *Mon. Not. R. Astr. Soc.* 203 (1983) 31–45.

Elmegreen, Bruce G., and Charles J. Lada. "Discovery of an extended (85 pc) molecular cloud associated with the M17 star-forming complex." *The Astronomical Journal* 81 (December 1976) 1089–1094.

——— "Sequential Formation of Subgroups in OB Associations." *The Astrophysical Journal* 214 (June 15, 1977) 725–741.

Evans, N. J., II. "Star Formation in Molecular Clouds," *Protostars and Planets.* Tucson: The University of Arizona Press, 1978, pp. 153–163.

Genzel, R., M. J. Reid, J. M. Moran, and D. Downes. "Proper Motions and Distances of H_2O Maser Sources. I. The Outflow in Orion-KL." *The Astrophysical Journal* 244 (March 15, 1981) 884–902.

Herbst, William, and George E. Assousa. "Supernovas and Star Formation." *Scientific American* (August 1979) 138–144.

Hoyle, Fred. *Frontiers of Astronomy.* New York: Harper & Row, 1955.

Lada, Charles J. "Energetic Outflows from Young Stars." *Scientific American* (July 1982) 82–93.

Lada, Charles J., Leo Blitz, and Bruce G. Elmegreen. "Star Formation in OB Associations," *Protostars and Planets.* Tucson: The University of Arizona Press, 1978, pp. 341–365.

Myers, P. C. "Low-Mass Star Formation in the Dense Interior of Barnard 18." *The Astrophysical Journal* 257 (June 15, 1982) 620–632.

Neugebauer, G., et al. "Early Results from the Infrared Astronomical Satellite." *Science* 224 (April 6, 1984) 14–21.

Robinson, Leif J. "The Frigid World of IRAS—I." *Sky & Telescope* (January 1984) 4–8.

Schorn, Ronald A. "The Frigid World of IRAS—II." *Sky & Telescope* (February 1984) 119–124.

Schramm, David N., and Robert N. Clayton. "Did a Supernova Trigger the Formation of the Solar System?" *Scientific American* (October 1978) 124–139.

Schwarzschild, Bertram M. "Are molecular clouds the heaviest objects in our Galaxy?" *Physics Today* (March 1980) 17–20.

——— "Infrared evidence for protoplanetary rings around seven stars." *Physics Today* (May 1984) 17–20.

Scoville, Nick, and Judith S. Young. "Molecular Clouds, Star Formation and Galactic Structure." *Scientific American* (April 1984) 42–53.

Shipman, Harry L. *The Restless Universe: An Introduction to Astronomy.* Boston: Houghton Mifflin Company, 1978.

Shklovskii, Iosif S. *Stars: Their Birth, Life and Death.* San Francisco: W. H. Freeman and Company, 1978.

Thaddeus, P. "Molecular Clouds in Orion and Monoceros." *Annals of the New York Academy of Sciences* (1982) 9–16.

——— "Radio observations of molecules in the interstellar gas." *Phil. Trans. R. Soc. Lond. A*. 303 (1981) 469–486.

Tucker, K. D., M. L. Kutner, and P. Thaddeus. "A Large Carbon Monoxide Cloud in Orion." *The Astrophysical Journal* 186 (November 15, 1973) L13–L17.

Turner, Barry E. "Interstellar Molecules." *Scientific American* (March 1973) 51–69.

Waldrop, M. Mitchell. "The Hunter and the Starcloud." *Science* 215 (February 5, 1982) 647–650.

——— "IRAS." *Science* 225 (July 6, 1984) 38–39.

Waldrop, M. Mitchell, and Richard A. Kerr. "IRAS Science Briefing." *Science* 222 (November 25, 1983) 916–917.

Whitney, Charles A. *The Discovery of Our Galaxy.* New York: Alfred A. Knopf, 1971.

Wynn-Williams, Gareth. "The Newest Stars in Orion." *Scientific American* (August 1981) 46–55.

Zuckerman, B. "A Model of the Orion Nebula." *The Astrophysical Journal* 183 (August 1, 1973) 863–869.

• CHAPTER TWO •

Bethe, Hans A., and Gerald Brown. "How a Supernova Explodes." *Scientific American* (May 1985) 60–68.

Burnell, Jocelyn Bell. "The Discovery of Pulsars." *Serendipitous Discoveries in*

Radio Astronomy, Proceedings of Workshop Number 7 held at the National Radio Astronomy Observatory, Green Bank, West Virginia, May 4–6, 1983.

Chevalier, Roger A. "Supernova Remnants." *American Scientist* 66 (November/December 1978) 712–717.

Fowler, William A. "The Quest for the Origin of the Elements." *Science* 226 (November 23, 1984) 922–935.

Greenstein, George. *Frozen Star.* New York: Freundlich Books, 1983.

Grindlay, Jonathan. "New Bursts in Astronomy." *Mercury* (September/October 1977) 6–11.

——— "An X-Ray Portrait of Our Galaxy." *Astronomy from Space.* Cambridge: Smithsonian Astrophysical Observatory, 1983, pp. 141–169.

Harnden, F. R., Jr. "X-ray telescope in space reveals unseen universe." *Smithsonian* 11 (September 1980) 110–114.

Helfand, David J. "Pulsars: Physics Laboratories in Our Galaxy." *Mercury* (May/June 1977) 2–7.

Kirshner, Robert P., and Roger A. Chevalier. "Spectra of Cassiopeia A: I. Observations." *The Astrophysical Journal* 218 (November 15, 1977) 142–147.

Kwok, Sun. "Not with a Bang but a Whimper." *Sky & Telescope* (May 1982) 449–455.

Mitton, Simon. *Daytime Star.* New York: Charles Scribner's Sons, 1981.

Paczynski, Bohdan. "Binary Stars." *Science* 225 (July 20, 1984) 275–280.

Reddy, Francis. "Supernovae: Still a Challenge." *Sky & Telescope* (December 1983) 485–490.

Schwarzschild, Bertram M. "Newly discovered pulsar is 20 times faster than Crab pulsar." *Physics Today* (March 1983) 19–21.

Sullivan, Walter. *Black Holes: The Edge of Space, The End of Time.* New York: Anchor Press/Doubleday, 1979.

Trimble, Virginia. "A Field Guide to Close Binary Stars." *Sky & Telescope* (October 1984) 306–311.

Trimble, Virginia, and Judith G. Cohen. "Some Faint Stars and a Bright One." *Engineering & Science* XLVIII (March 1985) 11–14.

Wali, Kameshwar C. "Chandrasekhar v. Eddington—an unanticipated confrontation." *Physics Today* (October 1982) 33–40.

Weymann, Ray J. "Stellar Winds." *Scientific American* (August 1978) 44–52.

Williams, Robert E. "The Shells of Novas." *Scientific American* (April 1981) 120–128+.

· CHAPTER THREE ·

Bartusiak, Marcia. "Experimental Relativity: Its Day in the Sun." *Science News* 116 (August 25, 1979) 140–142.
——— "Sensing the Ripples in Space-Time." *Science 85* (April 1985) 58–65.
Calder, Nigel. *Einstein's Universe.* New York: The Viking Press, 1979.
Cowley, A. P., D. Crampton, J. B. Hutchings, R. Remillard, and J. E. Penfold. "Discovery of a Massive Unseen Star in LMC X-3." *The Astrophysical Journal* 272 (September 1, 1983) 118–122.
Drever, Ronald W. P. "The Search for Gravitational Waves." *Engineering & Science* (January 1983) 6–9.
Fisher, Arthur. "Inventing the Wave Catchers." *Mosaic* (March/April 1983) 16–23.
Hawking, S. W. "The Quantum Mechanics of Black Holes." *Scientific American* (January 1977) 34–40.
Hoffmann, Banesh, and Helen Dukas. *Albert Einstein: Creator and Rebel.* New York: The Viking Press, 1972.
Hutchings, J. B. "Observational Evidence for Black Holes." *American Scientist* 73 (January/February 1985) 52–59.
Kaufmann, William J., III. *Black Holes and Warped Spacetime.* San Francisco: W. H. Freeman and Company, 1979.
Margon, Bruce. "The Bizarre Spectrum of SS 433." *Scientific American* (October 1980) 54–65.
Misner, Charles W., Kip S. Thorne, and John Archibald Wheeler. *Gravitation.* San Francisco: W. H. Freeman and Company, 1973.
Montagu, Ashley. "Conversations with Einstein." *Science Digest* (July 1985) 50ff.
Nicolson, Iain. *Gravity, Black Holes and the Universe.* New York: John Wiley & Sons, 1981.
Shapiro, Stuart L., Richard F. Stark, and Saul A. Teukolsky. "The Search for Gravitational Waves." *American Scientist* 73 (May/June 1985) 248–257.
Shipman, Harry L. *Black Holes, Quasars, and the Universe.* Boston: Houghton Mifflin Company, 1980.

Thorne, Kip S. "The Search for Black Holes." *Scientific American* (December 1974) 32–43.

Wheeler, John A. "Black Holes and New Physics." *Discovery* 7:2 (Winter 1982) 4–8.

· CHAPTER FOUR ·

Blitz, Leo. "Milky Way Spiral Structure: A New Look," *Surveys of the Southern Galaxy.* Boston: D. Reidel Publishing Company, 1983, pp. 117–125.

Blitz, Leo, Michel Fich, and Shrinivas Kulkarni. "The New Milky Way." *Science* 220 (June 17, 1983) 1233–1240.

Bok, Bart J. "The Milky Way Galaxy." *Scientific American* (March 1981) 92–120.

Cash, Webster, and Philip Charles. "Stalking the Cygnus Superbubble." *Sky & Telescope* (June 1980) 455–460.

Geballe, Thomas R. "The Central Parsec of the Galaxy." *Scientific American* (July 1979) 60–70.

Gingerich, Owen. "The Discovery of the Milky Way's Spiral Arms." *Sky & Telescope* (July 1984) 10–12.

Kuhn, Ludwig. *The Milky Way: The Structure and Development of our Star System.* New York: John Wiley & Sons, 1982.

Lebofsky, M. J., and G. H. Rieke. "M Supergiants and Star Formation at the Galactic Center." *The Astrophysical Journal* 263 (December 15, 1982) 736–740.

Lo, K. Y., and M. J. Claussen. "High-resolution observations of ionized gas in central 3 parsecs of the Galaxy: possible evidence for infall." *Nature* 306 (December 15, 1983) 647–651.

Morris, Mark. "Giant Magnetic Features Near the Core of Our Galaxy." *Physics Today* (January 1985) S7–S8.

Seiden, Philip E. and Humberto Gerola. "Propagating Star Formation and the Structure and Evolution of Galaxies." *Fundamentals of Cosmic Physics* Vol. 7 (1982) 241–311.

Smith, David H., and Leif J. Robinson. "Dissecting the Hub of Our Galaxy." *Sky & Telescope* (December 1984) 494–497.

Sofue, Yoshiaki, and Toshihiro Handa. "A radio lobe over the galactic centre." *Nature* 310 (August 16, 1984) 568–569.

Waldrop, M. Mitchell. "The Core of the Milky Way." *Science* 230 (October 11, 1985) 158–160.

Whitney, Charles A. *The Discovery of Our Galaxy.* New York: Alfred A. Knopf, 1971.

Wynn-Williams, Gareth. "The Center of Our Galaxy." *Mercury* (September/October 1979) 97–100.

· CHAPTER FIVE ·

de Vaucouleurs, Gerard. "The Distance Scale of the Universe." *Sky & Telescope* (December 1983) 511–516.

Dressler, A. "The Evolution of Galaxies in Clusters." *Annual Review of Astronomy and Astrophysics* 22 (1984) 185–222.

Gallagher, J. S., G. R. Knapp, and S. M. Faber. "H I Observations of Strongly Interacting Galaxies." *The Astronomical Journal* 86 (December 1981) 1781–1790.

Hodge, Paul. "The Cosmic Distance Scale." *American Scientist* 72 (September/October 1984) 474–482.

Hubble, Edwin. *The Realm of the Nebulae.* New Haven: Yale University Press, 1936.

Islam, J. N. *The Ultimate Fate of the Universe*. Cambridge: Cambridge University Press, 1983.

Jones, C., and W. Forman. "The Structure of Clusters of Galaxies Observed with *Einstein.*" *The Astrophysical Journal* 276 (January 1, 1984) 38–55.

Koo, David C. "Multicolor Photometry of the Red Cluster 0016+16 at z = 0.54." *The Astrophysical Journal* 251 (December 15, 1981) L75–L79.

Kron, Richard G. "The Evolution of Galaxies: Expectations and Observations." *Vistas in Astronomy* 26 (1982) 37–65.

——— "The Most Distant Known Galaxies." *Science* 216 (April 16, 1982) 265–269.

Larson, Richard B. "The Formation of Galaxies." *Mercury* (May/June 1979) 53–56.

Ostriker, Jeremiah P. "Galaxies: Outstanding Problems and Instrumental Prospects for the Coming Decade (A Review)." *Proceedings of the National Academy of Sciences USA* 74:5 (May 1977) 1767–1774.

Ostriker, J. P., and Scott D. Tremaine. "Another Evolutionary Correction to the Luminosity of Giant Galaxies." *The Astrophysical Journal* 202 (December 15, 1975) L113–L117.

Rieke, G. H., M. J. Lebofsky, R. I. Thompson, F. J. Low, and A. T. Tokunaga. "The Nature of the Nuclear Sources in M82 and NGC 253." *The Astrophysical Journal* 238 (May 15, 1980) 24–40.

Schweizer, François. "Colliding and Merging Galaxies. I: Evidence For the Recent Merging of Two Disk Galaxies in NGC 7252." *The Astrophysical Journal* 252 (January 15, 1982) 455–460.

Strom, K.M., and S.E. Strom. "Galactic Evolution: A Survey of Recent Progress." *Science* 216 (May 7, 1982) 571–580.

Toomre, Alar. "Interacting Systems," *The Large Scale Structure of the Universe: IAU Symposium No. 79.* Boston: D. Reidel Publishing, 1978, pp. 109–116.

Toomre, Alar, and Juri Toomre. "Violent Tides Between Galaxies." *Scientific American* (December 1973) 39–48.

· CHAPTER SIX ·

Bartusiak, Marcia. "Very Large Astronomy." *Science 84* (July/August 1984) 64–71.

Blandford, R. D. "Models of Active Galactic Nuclei." Eleventh Texas Symposium on Relativistic Astrophysics, *Annals of the New York Academy of Sciences,* Vol. 422 (March 1984).

Blandford, Roger D., Mitchell C. Begelman, and Martin J. Rees. "Cosmic Jets." *Scientific American* (May 1982) 124–142.

Burns, Jack O. and R. Marcus Price. "Centaurus A: The Nearest Active Galaxy." *Scientific American* (November 1983) 56–66.

Edelson, Edward. "Faster Than the Speed of Light?" *Mosaic* 13 (July/August 1982) 25–29.

Green, Richard F. and H. K. C. Yee. "An Imaging Survey of Fields Around Quasars. I: A Catalog." *The Astrophysical Journal Supplement Series* 54 (April 1984) 495–512.

Kennicutt, Robert C., Jr. and William C. Keel. "Induced Nuclear Emission-Line Activity in Interacting Spiral Galaxies." *The Astrophysical Journal* 279 (April 1, 1984) L5–L9.

Kraus, John. "The First 50 Years of Radio Astronomy, Part 2: Grote Reber and the First Radio Maps of the Sky." *Cosmic Search* 4:1 (1982) 14–18.

Oke, J. B. "Optical Variations in the Radio Galaxy 3C 371." *The Astrophysical Journal* 150 (October 1967) L5–L8.

Osmer, Patrick S. "Quasar as Probes of the Distant and Early Universe." *Scientific American* (February 1982) 126–138.

Overbye, Dennis. "Exploring the Edge of the Universe." *Discover* (December 1982) 22–28.

Readhead, Anthony C. S. "Radio Astronomy by Very-Long-Baseline Interferometry." *Scientific American* (June 1982) 51–61.

Schmidt, M. "3C 273: A Star-like Object with Large Red-Shift." *Nature* 197: 4872 (March 16, 1963) 1040–1042.

Schmidt, Maarten and Richard F. Green. "Quasar Evolution Derived from the Palomar Bright Quasar Survey and Other Complete Quasar Surveys." *The Astrophysical Journal* 269 (June 15, 1983) 352–374.

Schwarzschild, Bertram M. "Companion galaxies match quasar redshifts: the debate goes on." *Physics Today* (December 1984) 17–19.

Stockton, Alan, and John W. MacKenty. "3CR249.1 and Ton202—luminous QSOs in interacting systems." *Nature* 305 (October 20, 1983) 678–682.

Unwin, S. C., et al. "Superluminal Motion in the Quasar 3C 345." *The Astrophysical Journal* 271 (August 15, 1983) 536–550.

Waldrop, M. Mitchell. "NGC 4151: The Monster in the Middle." *Science* (December 2, 1983) 1003.

Wyckoff, Susan and Peter A. Wehinger. "Are Quasars Luminous Nuclei of Galaxies?" *Sky & Telescope* (March 1981) 200–204.

Wyckoff, S., P. A. Wehinger, and T. Gehren. "Resolution of Quasar Images." *The Astrophysical Journal* 247 (August 1, 1981) 750–761.

Wyckoff, S., et al. "Discovery of Nebulosity Associated with the Quasar 3C 273." *The Astrophysical Journal* 242 (December 1, 1980) L59–L63.

Yee, H. K. C., and Richard F. Green. "An Imaging Survey of Fields Around Quasars. II: The Association of Galaxies with Quasars." *The Astrophysical Journal* 280 (May 1, 1984) 79–90.

· CHAPTER SEVEN ·

Bartusiak, Marcia. "The Universe, By and Large." *Mosaic* 15:2 (1984) 2–9.

Chincarini, Guido, and Herbert J. Rood. "The Cosmic Tapestry." *Sky & Telescope* (May 1980) 364–371.

BIBLIOGRAPHY

Cornell, James. *The First Stargazers.* New York: Charles Scribner's Sons, 1981.

de Vaucouleurs, Gerard. "Supergalaxy." *Discovery* 5:2 (December 1980) 20–23.

——— "The Local Supercluster of Galaxies." *Bulletin of the Astronomical Society of India* 9 (1981) 1–23.

Geller, Margaret J. "Mapping the Universe." *World Book Science Year* (1984).

Gregory, Stephen A. and Laird A. Thompson. "Superclusters and Voids in the Distribution of Galaxies." *Scientific American* (March 1982) 106–114.

Gregory, Stephen A., Laird A. Thompson, and William G. Tifft. "The Perseus Supercluster." *The Astrophysical Journal* 243 (January 15, 1981) 411–426.

Helfand, David J. "Superclusters and the Large-Scale Structure of the Universe," *Physics Today* (October 1983) 17–20.

Kirshner, Robert P. "Giant Voids in the Universe." *1985 Yearbook of Science and the Future.* Chicago: Encyclopaedia Britannica, Inc., 1984.

——— "Hole in Space." *Natural History* (October 1982) 26–28.

Kirshner, Robert P., Augustus Oemler, Jr., Paul L. Schechter, and Stephen A. Shectman. "A Million Cubic Megaparsec Void in Boötes?" *The Astrophysical Journal* 248 (September 1, 1981) L57–L60.

Longair, M. S., and J. Einasto (editors). *The Large Scale Structure of the Universe: IAU Symposium No. 79.* Boston: D. Reidel Publishing Company, 1978.

Oort, J. H. "Superclusters." *The Annual Review of Astronomy and Astrophysics* 21 (1983) 373–428.

Ostriker, Jeremiah P. and Lennox L. Cowie. "Galaxy Formation in an Intergalactic Medium Dominated by Explosions." *The Astrophysical Journal* 243 (February 1, 1981) L127–L131.

Peebles, P. J. E. "The Origin of Galaxies and Clusters of Galaxies." *Science* 224 (June 29, 1984) 1385–1391.

Press, Frank and Raymond Siever. *Earth.* San Francisco: W. H. Freeman and Company, 1974.

Shane, C. Donald. "After 1945," Chapter 8. C. Donald Shane's handwritten autobiography on file in Mary Lea Shane Archives, University of California, Santa Cruz (May 1981).

Shane, C. D., and C. A. Wirtanen, "The Distribution of Extragalactic Nebulae." *Lick Observatory Bulletin* 528 (1954) 91–110.

Tully, R. Brent. "The Local Supercluster." *The Astrophysical Journal* 257 (June 15, 1982) 389–422.

Waldrop, M. Mitchell. "A Flower in Virgo." *Science* 215 (February 19, 1982) 953–955.

——— "The Large-Scale Structure of the Universe." *Science* 219 (March 4, 1983) 1050–1052.

Zel'dovich, Y. B., J. Einasto, and S. F. Shandarin. "Giant Voids in the Universe." *Nature* 300 (December 2, 1982) 407–413.

· CHAPTER EIGHT ·

Aaronson, Marc. "Accurate Radial Velocities for Carbon Stars in Draco and Ursa Minor: The First Hint of a Dwarf Spheroidal Mass-To-Light Ratio." *The Astrophysical Journal* 266 (March 1, 1983) L11–L15.

Blumenthal, George R., S. M. Faber, Joel R. Primack, and Martin J. Rees. "Formation of Galaxies and Large Scale Structure with Cold Dark Matter." *Nature* 311 (1984) 517.

Centrella, Joan and Adrian L. Melott. "Three-dimensional simulation of large-scale structure in the Universe." *Nature* 305 (September 15, 1983) 196–198.

Davis, Marc, George Efstathiou, Carlos S. Frenk, and Simon D. M. White. "The Evolution of Large-Scale Structure in a Universe Dominated by Cold Dark Matter." *The Astrophysical Journal* 292 (May 15, 1985) 371–394.

Gott, J. Richard, III. "Gravitational Lenses." *American Scientist* 71 (March/April 1983) 150–157.

Hegyi, Dennis J., and Keith A. Olive. "Can Galactic Halos Be Made of Baryons?" *Physics Letters* 126B (1983) 28–32.

Morrison, Nancy D., and David Morrison. "Unseen Mass in the Giant Elliptical Galaxy M87." *Mercury* (July/August 1983) 123–125.

Peebles, P. J. E. "The Origin of Galaxies and Clusters of Galaxies." *Science* 224 (June 29, 1984) 1385–1391.

Rubin, Vera C. "Dark Matter in Spiral Galaxies." *Scientific American* (June 1983) 96–108.

——— "The Rotation of Spiral Galaxies." *Science* 220 (June 24, 1983) 1339–1344.

Silbar, Margaret L. "The Lightweight Neutrino Weighs In." *Mosaic* (November/December 1980) 22–27.

Thomsen, Dietrick E. "A Closed Universe May Be Axiomatic." *Science News* 125 (June 23, 1984) 396–397.

Tucker, Wallace. "The Missing Mass Mystery." *Science Digest* (September 1981) 18.

Turner, Edwin L. "Gravitational Lenses and Particle Properties." Proceedings of the 1984 Fermilab Inner Space/Outer Space Conference.

——— "Information on Galaxy Clustering from Gravitational Lenses," *Clusters and Groups of Clusters.* Boston: D. Reidel Publishing Company, 1984, pp. 553–558.

Waldrop, M. Mitchell. "Massive Neutrinos: Masters of the Universe?" *Science* 211 (January 30, 1981) 470–472.

——— "Supersymmetry and Supergravity." *Science* 220 (April 29, 1983) 491–493.

White, Simon D. "The Evolution of Large-Scale Structure." Proceedings of the 1984 Fermilab Inner Space/Outer Space Conference.

· CHAPTER NINE ·

Davies, Paul. "Relics of Creation." *Sky & Telescope* (February 1985) 112–115.

Gamow, George. *My World Line.* New York: The Viking Press, 1970.

Guillen, Michael. "The Paradox of Antimatter." *Science Digest* (February 1985) 33–37+.

Hawking, Stephen. "The Goal of Theoretical Physics—1." *CERN Courier* (January/February 1981) 3–8.

———"The Goal of Theoretical Physics—2." *CERN Courier* (March 1981) 71–74.

Kolb, Edward W., and Michael S. Turner. "The Early Universe." *Nature* 294 (December 10, 1981) 521–526.

Lake, George. "Windows on a New Cosmology." *Science* (May 18, 1984) 675–681.

Lemaître, G. *The Primeval Atom.* New York: Van Nostrand, 1951.

Lin, Leslie. "Can We Understand the Universe?" *The Research News: University of Michigan* XXXII (June/July, 1981).

Mann, Charles C., and Robert P. Crease. "Waiting for Decay." *Science 86* (March 1986) 20–31.

Quigg, Chris. "Elementary Particles and Forces." *Scientific American* 252 (April 1985) 84–95.

Rees, Martin. "Our Universe—and Others." *New Scientist* (January 29, 1981) 270–273.

Schechter, Bruce. "The Moment of Creation." *Discover* (April 1983) 18–25.

Schramm, David N. "The Early Universe and High-Energy Physics." *Physics Today* (April 1983) 27–33.

Silbar, Margaret. "Gravity, the Fourth Force." *Mosaic* 15:2 (1984) 23–27.

Stecker, F. W. "Grand Unification and Possible Matter-Antimatter Domain Structure in the Universe." *Annals of the New York Academy of Sciences* (1981) 69–89.

Sulak, Lawrence R. "Waiting for the Proton to Decay." *American Scientist* (November/December 1982) 616–625.

Taubes, Gary. "Detecting Next to Nothing." *Science 85* (May 1985) 58–66.

Thomsen, D. E. "Gathering String in the Cosmos." *Science News* 125 (May 12, 1984) 294–295.

Wagoner, Robert V., and Donald W. Goldsmith. *Cosmic Horizons*. San Francisco: W. H. Freeman and Company, 1983.

Waldrop, M. Mitchell. "Matter, matter, everywhere. . . ." *Science* 211 (February 20, 1981) 803–806.

Weisskopf, Victor F. "The Origin of the Universe." *American Scientist* 71 (September/October 1983) 473–480.

Wilczek, Frank. "The Cosmic Asymmetry Between Matter and Antimatter." *Scientific American* (December 1980) 82–90.

• CHAPTER TEN •

Albrecht, Andreas, and Paul J. Steinhardt. "Cosmology for Grand Unified Theories with Radiatively Induced Symmetry Breaking." *Physical Review Letters* 48 (April 26, 1982) 1220–1223.

Guth, Alan H. "Inflationary universe: A possible solution to the horizon and flatness problems." *Physical Review* D23 (January 15, 1981) 347–356.

——— "Speculations on the Origin of the Matter, Energy, and Entropy of the Universe," *Asymptotic Realms of Physics: Essays in Honor of Francis E. Low*. Cambridge: MIT Press, 1982, pp. 199–222.

Guth, Alan H., and Paul J. Steinhardt. "The Inflationary Universe." *Scientific American* 250 (May 1984) 116–128.

Kazanas, Demosthenes. "Dynamics of the Universe and Spontaneous Symmetry Breaking." *The Astrophysical Journal* 241 (October 15, 1980) L59–L63.

Levi, Barbara G. "New Inflationary Universe: An Alternative to Big Bang?" *Physics Today* 36 (May 1983) 17–19.

Overbye, Dennis. "The Universe According to Guth." *Discover* (June 1983) 93–99.

Smith, David H. "The Inflationary Universe Lives?" *Sky & Telescope* (March 1983) 207–210.

Trefil, James S. *The Moment of Creation.* New York: Charles Scribner's Sons, 1983.

Waldrop, M. Mitchell. "Before the Beginning." *Science 84* (January/February 1984) 45–51.

——— "In Search of the Magnetic Monopole." *Science* 216 (June 4, 1982) 1086–1088.

——— "The New Inflationary Universe." *Science* 219 (January 28, 1983) 375–377.

· CHAPTER ELEVEN ·

Atkatz, D., and H. Pagels. "Origin of the Universe as a Quantum Tunneling Event." *Physical Review D* 25:8 (April 15, 1982) 2065–2073.

Barnes, Harry Elmer. *An Intellectual and Cultural History of the Western World,* vol. 2. New York: Dover Publications, 1965.

Brout, R., F. Englert, and E. Gunzig. "The Causal Universe." *General Relativity and Gravitation* 10:1 (January 1979) 1–6.

Davies, P. C. W. *Space and Time in the Modern Universe.* Cambridge: Cambridge University Press, 1977.

DeWitt, Bryce S. "Quantum Gravity." *Scientific American* (December 1983) 112–129.

Freedman, Daniel Z., and Peter van Nieuwenhuizen. "The Hidden Dimensions of Spacetime." *Scientific American* (March 1985) 74–81.

Gale, George. "The Anthropic Principle." *Scientific American* (December 1981) 154–171.

Gott, J. Richard, III. "Creation of Open Universes from de Sitter Space." *Nature* 295 (January 28, 1982) 304–307.

Gott, J. Richard, III, and Thomas S. Statler. "Constraints on the Formation of Bubble Universes." Princeton Observatory Preprint, July 11, 1983.

Guillemin, Victor. *The Story of Quantum Mechanics.* New York: Charles Scribner's Sons, 1968.

Hawking, S. W. "The Edge of Spacetime." *American Scientist* 72 (July/August 1984) 355–359.

Loudon, Jim. "Cosmic dice." *Science News* 128 (September 14, 1985) 163.

MacRobert, Alan. "Beyond the Big Bang." *Sky and Telescope* (March 1983) 211–213.

Misner, Charles, Kip S. Thorne, and John Archibald Wheeler. *Gravitation.* San Francisco: W. H. Freeman and Company, 1973.

Pagels, Heinz R. "Before the Big Bang." *Natural History* (April 1983) 22–26.

Schwarz, John. "Completing Einstein." *Science 85* (November 1985) 60–64.

Thomsen, Dietrick E. "Kaluza-Klein: The Koenigsberg Connection." *Science News* 126 (July 7, 1984) 12–14.

Tryon, Edward P. "Is the Universe a Vacuum Fluctuation?" *Nature* 246 (December 14, 1973) 396–397.

Vilenkin, Alexander. "Creation of Universes from Nothing." *Physics Letters* 117B (November 4, 1982) 25–28.

Waldrop, M. Mitchell. "Bubbles Upon the River of Time." *Science* 215 (February 26, 1982) 1082–1083.

Wow-Mann, G. *Superduperstuff in the Universe*. Venice Beach, California, The Universe: Institute for Innerspace/Outerspace Interface, 1984.

INDEX

INDEX

INDEX

INDEX

INDEX

INDEX

INDEX

INDEX

INDEX

INDEX

Photo of Marcia Bartusiak by Karen Keeney, *Discover Magazine*

About the Author

Award-winning writer Marcia Bartusiak holds a B.A. in communications from American University and an M.S. in physics from Old Dominion University. Combining her communications skills with a science background, Bartusiak has written on a variety of subjects—including relativity, plate tectonics, astrochemistry, and computer graphics—for *Omni, Discover, Popular Science, Science News, Air & Space,* the *New York Times Book Review, Sky & Telescope,* and *Mosaic,* the National Science Foundation magazine. She is a recipient of the American Institute of Physics Science Writing Award for her articles on atomic clocks—the first woman to win the award since it was established in 1968. THURSDAY'S UNIVERSE was honored as one of the Astronomy Books of the Year for 1987 by the Astronomical Society of the Pacific. Bartusiak lives in Norfolk, Virginia.

Other Titles from Tempus Books

The World of Mathematics

A Small Library of the Literature of Mathematics from A'h-mosé the Scribe to Albert Einstein

Published in 1956 to critical acclaim and with sales of more than 200,000 copies, THE WORLD OF MATHEMATICS is now available for a new generation of readers. Out of print for many years, this four-volume anthology is a rich collection of 133 articles. The essays provide first-person windows on the concepts of pure math, the laws of probability, statistics, puzzles and paradoxes, and the role of mathematics in economics, art, music, and literature. Each article is prefaced by Newman's insightful commentary that places it in historical perspective and makes even the most difficult concepts accessible to a wide range of readers. The selections range from hard-to-find classical pieces by Archimedes, Galileo, and Mendel to works of twentieth-century thinkers, including John Maynard Keynes, Bertrand Russell, and A.M. Turing.

James R. Newman, 2800 pages (four volumes), $50.00 softcover, #86-96544, $99.95 cloth, #86-96593

Invisible Frontiers

The Race to Synthesize a Human Gene

"An important and pioneering book, dealing with events of high scientific and economic consequence.... [Hall] succeeds marvelously in making the science accessible to the general reader."

New York Times Book Review

INVISIBLE FRONTIERS tracks the developments of the high-stakes race to clone a human gene and engineer the mass production of the life-sustaining hormone insulin. Stephen Hall takes the reader behind the scenes to the labs of the three main players: Harvard University, the University of California at San Francisco, and City of Hope National Medical Center in the Los Angeles area. Hall weaves together the threads of the story—the brilliance of modern-day research, the combativeness of the labs, the clashing egos of scientists, the incursion of commercial interests, the persistent presence of regulation, and the unpredictable effects of local politics. An authoritative and vivid account.

Stephen S. Hall, 322 pages, $8.95 softcover, #86-96817

The New Wizard War

How the Soviets Steal U.S. High Technology — And How We Give It Away

Is the United States giving the Soviet Union an unprecedented technological edge in the name of free trade? How easy is it for Soviet spies to smuggle high-technology equipment and information to the Soviet Union? How adept is the U.S. government at controlling our technology exports? And are our allies abetting the Soviet acquisition enterprise? THE NEW WIZARD WAR is a riveting and timely look at the legal and illegal transfer of high technology from the U.S. to the

U.S.S.R. The case studies in the book—mixing international intrigue, military secrets, million-dollar payoffs, and bureaucratic snafus—read like scenes from a fast-paced spy thriller. Much more than a good read, THE NEW WIZARD WAR is an eye-opening, provocative report on controversial and urgent election-year issues: the loss of U.S. technological leadership, the question of national security, the wisdom of U.S.-Soviet scientific cooperation, Japan's role in technology transfer, the Gorbachev agenda, and the balance between government and private control over the flow of technology.

Robyn Shotwell Metcalfe, 256 pages, $17.95 cloth, #86-96338

Time

The Familiar Stranger

"Fraser is perhaps the leading intellectual authority in the world on the study of time. Now he has given us a book that is both stimulating and provocative."

Jeremy Rifkin, author of *Time Wars*

This wide-ranging and learned book surveys the enormous variety of our understandings of time, both in the everyday world and in the sciences and humanities. From the visions of time and the timeless in religion, to those in contemporary physics and cosmology and the more common conceptions of time, J.T. Fraser offers the general reader a fascinating history of the idea and experience of time. Included are such issues as the beginning and end of the universe; the biology of aging and death; human perception of time; dreaming *vs* waking reality; expectation and memory; calendars and chronologies; and the ways in which technological rhythms control our lives.

J.T. Fraser, 400 pages, $9.95 softcover, #86-96809
Available in November 1988

Mathematics

Queen and Servant of Science

"Bell is a lively, stimulating writer, with a good sense of historical circumstance... a sound grasp of the entire mathematical scene, and a gift for clear and orderly explanation."

James R. Newman, editor of *THE WORLD OF MATHEMATICS*

Here—from a modern-day master of mathematics literature—is a fascinating and lively survey of the developments of pure and applied mathematics. E.T. Bell explores mathematical theories from those of Pythagoras and Euclid to those of Einstein and Gödel with vigor, intelligence, and wit that are unsurpassed. He covers a broad range of traditional subjects—algebra, geometry, calculus, topology, logic, rings, and groups—making even the most abstruse subjects come alive. Published in cooperation with the Mathematical Association of America.

Eric Temple Bell, 432 pages, $11.95 softcover, #86-96825
Available in November 1988

The Tomorrow Makers

A Brave New World of Living-Brain Machines

"Talespinners have nothing on the hard-core science freaks. This nonfiction book has enough new ideas for 16 Star Trek sequels. And better dialogue."

Rudy Rucker, *San Francisco Chronicle,* author of *Infinity and the Mind*

THE TOMORROW MAKERS is a spellbinding account of visionary researchers and scientists and their modern-day quest to engineer immortality. Award-winning science writer Grant Fjermedal details the astounding work being done in robotics and artificial intelligence today—on organic molecular computers smaller than a grain of sand; on lifelike androids that are intelligent and charming; and—most extraordinary of all—on the process of downloading human minds into computerized robots that will live forever. Cited by the American Library Association as one of 1987's most notable books.

Grant Fjermedal, 288 pages, $8.95 softcover, #86-96247

Nobel Dreams

Power, Deceit, and the Ultimate Experiment

"...one of those rare science books that tell about science in the course of telling about the human comedy."

Lee Dembart, *Los Angeles Times*

An intimate and colorful look at Carlo Rubbia's quest for the 1985 Nobel Prize in physics. It is a vivid saga of big science, big ego, frenetic competition, and hard-ball politics that demonstrates the chaotic and chancy nature of scientific research. Taubes' narrative is a very readable layperson's primer on high-energy physics, as well as a captivating modern-day adventure story.

Gary Taubes, 288 pages, $8.95 softcover, #86-96239

Inventors at Work

Interviews with 16 Notable American Inventors

"Invention becomes an art in these accounts of serendipitous associations and 'lateral thinking.'"

Publishers Weekly

A critically acclaimed collection of 16 engaging and illuminating interviews with the most notable inventors of our time—from professional R and D specialists such as NASA's Maxime Faget, Hughes Aircraft's Harold Rosen, and Xerox's Bob Gundlach to independents such as entrepreneur Stanford Ovshinsky and artificial intelligence expert Raymond Kurzweil. Their accomplishments—the laser, the microprocessor, the man-powered airplane, the implantable pacemaker, the Apple II computer, the plastic soda bottle, and many others—represent both the boldly significant and the subtly brilliant. Along with fascinating individual stories, INVENTORS AT WORK reveals instructive glimpses into the creative process, thoughts on the personal challenges and institutional roadblocks an inventor faces today, a look at invention as art and invention by committee, and discussions of the impact of United States patent laws and a capitalist economy on the inventive spirit.

Kenneth A. Brown, 408 pages, $9.95 softcover, #86-96080, $17.95 cloth, #86-96312

Computer Lib/Dream Machines

"If you want a vision at once backwards towards the dawning of the personal computer age and forward beyond the limits of the machines we use today, take a look at COMPUTER LIB."

Jim Seymour, syndicated columnist

First published in 1974, COMPUTER LIB became the first cult book of the computer generation, predicting the major issues of today: design of easy-to-use computer systems, image synthesis, artificial intelligence, and computer-assisted instruction. Ted Nelson's vision of a nonsequential way of storing data—hypertext—is particularly relevant today with the emergence of CD ROM. Republished by Tempus Books, COMPUTER LIB's wildly Utopian introduction to computers is now available to a new generation. Included is new material from Ted Nelson—commentaries, insights, updates, and reconsiderations—all in his characteristically opinionated, uplifting, and irreverent style.

Theodor Nelson, 336 pages, $18.95 softcover, #86-96031

The Pursuit of Growth

The Challenges, Opportunities, and Dangers of Managing and Investing in Today's Economy

Success stories of phenomenal business growth—IBM, DuPont, Proctor and Gamble—have been well publicized, as have spectacular growth failures—Atari, Bendix, People Express. What is this peculiar obsession with growth at any cost? When does it serve the best interests of managers, entrepreneurs, politicians, and the community? And when is it a formula for disaster? The authors provide a highly readable and insightful analysis of growth as an underlying tenet of business and government.

G. Ray Funkhouser and Robert R. Rothberg, 274 pages, $17.95 hardcover, #86-96098

Machinery of the Mind

Inside the New Science of Artificial Intelligence

"An ideal presentation of what artificial intelligence is all about."

Douglas Hofstadter, author of *Gödel, Escher, Bach*

Focusing on the work of giants in the artificial intelligence field—including Marvin Minsky, Roger Schank, and Edward Feigenbaum— George Johnson gives us an intimate look at the state of AI today. We see how machines are beginning to understand English, discover scientific theories, and create original works of art; how research in AI is helping us to understand the human mind; and how AI affects our lives today. Captivating reading for anyone with an interest in science and technology.

George Johnson, 352 pages, $9.95 softcover, #86-96072

Front cover photo:

The Horsehead Nebula in Orion

This photograph, produced by combining three separate photographs taken with the 1.2m UK Schmidt Telescope through different colored filters, is probably the most effective picture yet obtained of this famous and much-photographed region of the sky. It reveals not only the striking Horsehead itself, but also the faint structure of the bright nebula NCG 2024, just below the brilliant star Zeta Orionis. To achieve this, the greatly overexposed image of the star has been reduced by a photographic masking technique. Embedded in the large dust cloud below the Horsehead are two luminous stars whose light is reflected by the dust to give rise to two bluish nebulae, NC 2023 (nearest to the Horsehead) and IC 435, below.

Back cover photo:

Star Trails around the South Celestial Pole

Photo by David Malin, Anglo-Australian Observatory